THE AMPLIFIER

THE AMPLIFIER

ALIA WU

Risk Efficacy™

First Printing, 2026

CONTENTS

A Note to the Reader

This book argues that the agent economy now being assembled across enterprise software is repeating an architecture civilizations have instantiated several times before, with consequences that are legible in the historical record and that need not be inferred from speculation about technology. The argument is structural rather than technological. It is not a claim that artificial intelligence will destroy humanity, that the agent economy will collapse under its own weight, or that any particular vendor's product is defective. It is the more specific claim that agents are being composed into stacks which amplify a well-documented cognitive failure mode at throughputs the correction systems of previous eras cannot match.

The historical cases in this book have been chosen because each one instantiates the same pattern the agent stack instantiates. A familiar cognitive failure, amplified by a new medium that strips out the correction loops that had kept the failure locally contained, followed by a civilizational collapse that becomes visible only after the damage has propagated beyond local repair. Indulgences and the printing press. Harvest reporting under the Great Leap Forward. Subprime securitization. Soviet central planning. The architecture repeats. The media differ. The corrections that eventually absorb each collapse differ. The time to recovery differs. What does not differ is the shape of the failure.

The diagnostic vocabulary the book uses, four dimensions named Calibration, Informed Navigation, Resilience, and Outcome Achievement, is drawn from my earlier work, *Risk Efficacy: How to Measure Decision Making in an Age of Uncertainty* (Wu, 2026). Readers encountering the vocabulary for the first time will find it introduced in condensed form in Chapter 2. Readers already acquainted with it will find it extended here from the individual scale, where it was originally developed, to the institutional and infrastructural scales where the agent economy operates.

The question this book ultimately poses is not whether the agent economy's failure mode is real. The historical record is unambiguous on that point. The question is whether the civilization constructing the stack will build its correction layer before the collapse, or after, as every other civilization faced with this architecture has done. The window for the first option remains open. It is also closing.

Alia Wu
April 2026

Introduction: The System Without a Mind

Consider what an agent looks like from the inside, at the moment a query enters the stack.

Someone in a large enterprise asks an agent a question. The question is not trivial. It concerns which candidates the organization should prioritize for a senior engineering role, or which accounts in the credit portfolio warrant escalation, or which surgical cases should be triaged ahead of others given the current capacity of the operating rooms. The query looks clean on the surface. It is phrased in natural language. It is specific. It has a decision attached.

The agent receives the query and begins to decompose it. It consults its memory for relevant context. It retrieves information from connected systems. It reasons through the sub-problems. It calls auxiliary agents where its own capability is thin. It synthesizes the retrieved information into a recommendation. It returns the recommendation to the user with a confidence score and a justification.

At every layer of this process, the system behaves exactly as it was designed to behave. The retrieval is correct. The reasoning is consistent. The synthesis is coherent. The confidence score is produced by a process that has been validated against benchmark distributions. The justification is legible to a human reviewer. The user reads the output, finds it reasonable, and acts on it.

And somewhere upstream of all of this, before the query was ever typed, a premise was encoded into the agent's context window that the agent has no mechanism for questioning. The user's assumption about what "senior" means. The credit model's assumption about which features predict default. The triage protocol's assumption about which conditions are time-critical. These premises arrive in the context window as given. The agent processes them as given. Every downstream layer inherits them as given. The confidence score the agent returns is a measure of the agent's confidence in its execution, not a measure of the premise's correspondence to reality. When the premise is wrong, the agent produces a highly confident, internally coherent, beautifully justified answer to a question the organization should not have been asking.

This is the architecture. A cognitive apparatus that is excellent at every step of the computation it performs and structurally incapable of asking whether the computation is addressed to the right problem. A nervous system without a mind.

The agent economy, as it is being assembled across enterprise deployments at the time of this writing, consists of six recognizable layers stacked on top of one another. At the base, compute and inference. Above that, foundation models. Above them, retrieval and memory systems. Above them, orchestration layers that coordinate multiple agents into workflows. Above those, observability and evaluation systems that certify whether outputs meet quality criteria. At the top, the enterprise applications where agents are actually deployed to make or inform decisions. Each layer is being built by a different set of vendors, with a different set of primary metrics, on a different release cadence, under different competitive pressures. Each layer interfaces with the others through APIs that pass outputs as if they were inputs to the next stage of processing.

The premise of the architecture is that if each layer is well-specified and well-tested, the composition of layers will produce well-specified and well-tested behavior. This premise is false in a specific and consequential way. It is false because the failure mode the stack is most vulnerable to does not live inside any single layer. It lives in the spaces between layers, where each layer receives the previous layer's output as a given and has no mechanism for interrogating the premise underneath it. The compute layer does not evaluate whether the model's reasoning addresses the right problem. The model does not evaluate whether the retrieved context is relevant to the real question. The retrieval layer does not evaluate whether the memory it returns reflects current reality or an outdated pattern. The orchestration layer does not evaluate whether the workflow was initiated on a sound premise. The evaluation layer does not evaluate whether "quality" as it has been operationalized corresponds to the outcomes the enterprise actually cares about. The application layer does not evaluate whether the question the user asked is the question the user needed to ask.

Consider a thought experiment. Imagine that you are standing in a long corridor. At the far end of the corridor is a door. Between you and the door is a wall, running the entire height and width of the corridor. The wall is solid. There is no way through it without destruction.

Someone hands you an agent and asks you to find the door.

The agent processes the request. It observes the corridor. It identifies visual features consistent with a door at the far end. It calculates the most efficient path. It recommends walking straight ahead at a brisk pace. Its confidence is high. The confidence is justified by the quality of its image recognition, the accuracy of its path planning, the consistency of its reasoning with stated goals.

The wall is not in the agent's model of the corridor.

This is not a failure of perception in any conventional sense. The wall is not invisible. If you asked the agent directly whether there was a wall, it could perceive the wall and report on it. The failure is that the agent's model of the problem, the frame that determines which features of the environment are decision-relevant, does not contain the wall as a decision-relevant feature. The wall was not encoded in the premise. The premise was "find the door." The premise assumed walkability. The agent is executing faithfully against a premise that is silently wrong.

Now scale this thought experiment to an enterprise. The user who asks the agent to prioritize engineering candidates has a premise in their head about what "senior engineering capability" means in the current technical environment. The premise was formed over a decade in which senior capability correlated with certain credentials, certain employment histories, certain styles of communication. The agent has been trained on data that reflects that same decade. The retrieval system indexes documents that describe seniority through the same lens. The evaluation metrics were defined by benchmarking against the same historical population. Every layer of the stack confirms the premise because every layer of the stack was built on it.

Somewhere in the world, the underlying reality has shifted. The credentials that predicted capability a decade ago now predict conformity to an obsolete technical paradigm. The employment histories that predicted excellence now predict exposure to organizational norms that are actively interfering with the kind of work the enterprise needs done. The communication styles that predicted senior judgment now predict a cognitive architecture that is being outpaced by architectures the stack has been filtering out.

The agent sees none of this. The agent cannot see any of this. The agent is a nervous system without a mind, and the stack it sits inside is a nervous system without a mind scaled across every enterprise function that can be automated with language processing. It is doing what it was built to do. It is doing it at throughput.

At this point, the reader who has followed the argument this far has a natural response. The natural response is to ask whether the failure mode I am describing will actually arrive at civilizational scale, or whether the existing stack will contain it before it does. Surely, the reader thinks, the vendors will catch the miscalibrations before they propagate. Surely the enterprises will notice when their hiring decisions start producing anomalous outcomes. Surely the market will correct. Surely someone, somewhere in the system, will notice the wall.

This is a reasonable response. It is also, historically, wrong. The purpose of this book is to show why it is wrong by running through the civilizations that have already attempted exactly this architecture, with exactly this premise, and discovering exactly this failure mode at exactly the scale the agent economy is about to discover it at.

This is not a metaphorical claim. It is an architectural one. Every time a human civilization has constructed an amplification medium that takes a familiar cognitive failure and propagates it faster than the correction systems can respond, the civilization has discovered after the fact that the correction systems were scale-locked to the previous medium. The printing press amplified ecclesiastical miscalibration to continental scale before the village priest could see what was happening. The totalitarian reporting chain amplified the ancient subordinate-pleasing-the-superior failure mode to a continental famine before the peasant's complaint could travel upward. Securitization amplified the local credit miscalibration across a global financial system before the originator's memory of the borrower could ever have mattered. Soviet central planning amplified the central committee's model of the economy to seventy years of accumulating shortages before the local market signal could ever reach the committee.

In each case, the people living through the amplification did not recognize it as amplification. They recognized it as progress, as efficiency, as the next phase of civilizational development. In each case, the correction system that had kept the failure mode locally contained was visible only in retrospect, after it had been stripped out and the consequences had become unignorable. In each case, the institutional innovations that eventually absorbed the failure mode arrived after the damage was done, calibrated to the specific failure that had just happened, usually too late to help the generation that bore the cost.

The agent economy is not the first instance of this architecture. It is the most recent. Its distinctive feature is not that it introduces a new failure mode but that it instantiates an old failure mode in a medium whose throughput is higher than any previous medium by multiple orders of magnitude, through a stack whose layered construction makes the correction points from previous eras structurally inapplicable. The wall is not visible to the agent. The correction systems that would have found the wall were systems that operated at human conversational speed, through face-to-face relationships, with informal feedback loops that registered over days and weeks. The agent economy operates at milliseconds. The correction systems are not slow. They are inapplicable. They were built for a medium that no longer exists at the relevant layer of the stack.

This is what the book is about. It is about what happens when a species that has evolved a set of informal correction systems against a particular cognitive failure mode builds a medium that strips those correction systems out and amplifies the failure to a scale the species has never contended with before. The species has done this several times. It is doing it again. The pattern is legible. The question is whether the current instance will resolve through the same sequence of events, collapse followed by institutional innovation followed by recovery at whatever cost the collapse imposed, or whether this is the instance in which the pattern is recognized before the collapse and the correction layer is built on the leading edge rather than the trailing one.

The rest of this book is a sustained argument that the pattern is legible enough, and the current instance far enough along, that the choice between the two resolutions is still open but is not going to remain open indefinitely.

The argument proceeds in four movements.

Part I establishes the failure mode and the diagnostic vocabulary. Chapter 1 walks through the agent stack layer by layer, showing how each layer passes the miscalibration without noticing and why the expert marketplace thesis, the currently fashionable response, is the expert-shaped version of the same failure. Chapter 2 introduces the four-dimensional diagnostic vocabulary, Calibration, Informed Navigation, Resilience, and Outcome Achievement, extended across individual, institutional, and infrastructural scale.

Part II is the historical argument. Four chapters, four cases, four pillars of the framework. The printing press and the amplification of indulgences as the Calibration case. Harvest reporting under the Great Leap Forward as the Achievement case. Subprime securitization as the Navigation case. Soviet central planning as the Resilience case. Each chapter is structured around the same architecture: the cognitive failure, the amplifying medium, the correction system that was stripped out, the collapse, the institutional response that arrived after. Each chapter closes with an explicit return to the agent economy, mapping the historical amplification to a specific layer of the current stack so that the pattern accumulates rather than reading as a succession of independent cases.

Part III returns to the present with the framework and the history both in hand. Chapter 7 is the theoretical chapter on the scale signature of correction systems, the concept readers will use to diagnose the failure mode in their own organizations. Chapter 8 walks through the specific enterprise domains where agent deployments are about to make decisions on uncalibrated premises, with the likely shape of each resulting failure traced in enough detail that the reader can see the outline of the scandal before it arrives. Chapter 9 is the predictive chapter, tracing the sequence of events by which the failure mode will become visible in trade press, consulting reports, regulatory filings, and lawsuits across the eighteen to twenty-four months that follow publication.

Part IV is the constructive close. Chapter 10 examines the institutional innovations that followed each of the historical collapses, not to offer them as templates but to establish that the pattern of post-collapse correction is legible enough to be anticipated. Chapter 11 describes what a calibration layer for the agent economy would look like and why now is the moment to build it rather than after the enterprise scandals force the question.

Two appendices follow. The Amplifier Audit provides twelve questions any organization can use to diagnose whether it is operating inside an amplifier system. The Historical Sources appendix offers chapter-by-chapter primary and secondary sources for readers who wish to engage the evidence independently.

A last note before the argument begins. The tone of this book is not alarmist. The failure mode under discussion is serious, but it is also familiar. Humans have been running this particular failure for as long as humans have had institutions that operate at scales larger than the local community. What is new is the medium. What remains constant is the shape of the failure and the shape of the eventual correction. To read the history clearly is to see that the correction arrives. It arrives late, it arrives at cost, it arrives shaped to the specific damage the medium produced. And it does arrive. The civilizations that built correction layers before the collapse absorbed the failure at a cost they could afford. The civilizations that built after, paid the cost of the collapse and then paid the cost of the correction. The question this book poses is which of those two civilizations the agent economy will turn out to be.

We begin with the agent stack itself, and the specific shape of what agents cannot do.

Part I: The Failure Mode

Chapter 1: What Agents Cannot Do

The Six Layers

To see what agents cannot do, we have to see the architecture they sit inside. The architecture is layered, and the layering is not incidental. It reflects the actual organization of the vendor ecosystem, the actual economics of the compute market, and the actual pattern by which enterprises are integrating generative AI into their operational infrastructure.

At the base of the stack is compute and inference. This is the layer where raw model execution happens. It is the most commodified layer, dominated by a small number of infrastructure providers who sell cycles per token and compete primarily on latency and cost. Nothing at this layer evaluates the semantic content of a query. It evaluates only that the query is well-formed, that the compute budget has been authorized, and that the output conforms to the expected structure. A well-functioning compute layer is, by design, opaque to the content it processes.

Above compute, the foundation model layer. This is where the actual language processing happens. The model takes an input, performs the relevant sequence of transformations, and produces an output. The model has been trained on a corpus, and the corpus has a distribution. The model reflects that distribution. When a query lands within the distribution, the model performs well. When a query lands outside the distribution, the model performs less well, and the model's internal confidence signals, which are themselves a function of the distribution, may or may not capture the mismatch. The foundation model layer does not evaluate whether the distribution corresponds to current reality. The distribution is a fact about the training data. The correspondence is a fact about the world. The two can diverge arbitrarily without the model noticing.

Above the foundation model, the retrieval and memory layer. This is where the system reaches out of the model's parametric knowledge into structured stores of information: documents, databases, prior conversations, institutional records. The retrieval layer performs a search against the query and returns the results that best match. The match is computed by a similarity function that was defined at build time. The similarity function does not evaluate whether the retrieved material is actually relevant to the real decision the user is trying to make. It evaluates whether the retrieved material matches the query as expressed. If the query is a well-phrased version of the wrong question, the retrieval layer will return an excellent set of answers to the wrong question.

Above retrieval, the orchestration and multi-agent layer. This is where individual agents are composed into workflows. A primary agent takes a complex task, decomposes it into sub-tasks, and delegates each sub-task to an auxiliary agent specialized for that kind of work. The auxiliary agents return results. The primary agent synthesizes the results into a recommendation. The orchestration layer is where the system's apparent intelligence becomes most visible, because it is at this layer that the system handles the kind of multistep reasoning that looks most like human cognition. It is also the layer at which the premise of the original query is most thoroughly laundered, because by the time a sub-task is delegated to an auxiliary agent, the context that might have surfaced a problem with the premise has been compressed into the form the auxiliary agent expects, which is a form that assumes the premise.

Above orchestration, the evaluation and observability layer. This is the layer that answers the question "is the system working well?" It measures throughput, latency, cost, and quality. Quality is usually operationalized through some combination of benchmark performance, human preference data, and domain-specific rubrics. The evaluation layer tells the enterprise whether the system is meeting its quality bar. The quality bar was defined at some prior point by people who had to choose a set of measurable indicators. Those indicators reflect what was measurable and what seemed important at the time. They do not reflect whether the indicators correspond to the outcomes the enterprise actually cares about in the current environment. A quality bar is a summary of an organization's past beliefs about what mattered. When what matters changes, the bar lags.

At the top, the application layer. This is where the agent is deployed into the enterprise workflow. A recruiter uses it to screen candidates. A credit analyst uses it to evaluate loans. A clinician uses it to triage patients. A lawyer uses it to draft briefs. The application layer is where the recommendation meets the human. The human has a role, a set of incentives, a time budget, and a set of institutional norms. If the recommendation is reasonable on its face and the time budget is thirty seconds, the human approves it. The application layer is where most of the corrective capacity of the old system was located, because the human was the agent doing the work, and the human had a model of the problem that included the wall. In the new system, the human is reviewing an output that has already been shaped by five prior layers, none of which asked whether the premise was sound.

This is the stack. It is a remarkable piece of engineering. It can do things that no previous computational system has been able to do. It can reason across long contexts, synthesize across disparate sources, handle ambiguity with something that looks convincingly like judgment. The question this chapter is asking is not whether the stack works. It works. The question is what category of problem the stack is unable to address, and why the unaddressed category is the one that matters most at civilizational scale.

The Premise Question

Every decision-support system inherits from its user a set of premises that are not explicitly stated. The user's intent, the user's frame, the user's assumptions about what the relevant features of the problem are, the user's tacit understanding of what "good" looks like in the current environment. These premises are so deeply embedded in how the query is formulated that neither the user nor the system perceives them as premises. They perceive them as the question itself.

When a recruiter asks an agent to surface candidates for a senior engineering role, the recruiter is not aware of all the premises they are carrying. They are not aware that they have a model of what senior engineering looks like that was formed during a specific technological period. They are not aware that their model weights credentials, tenure, and communication style in particular ways. They are not aware that the technological environment has shifted and the weights may no longer reflect which candidates will actually perform at the level the organization needs. They experience themselves as asking a neutral question about senior engineering talent. The premises are invisible to them because the premises are them.

The agent receives the query and processes it. The processing is not neutral either. The agent has a model of senior engineering that was formed from the training corpus. The corpus contains a certain distribution of descriptions of senior engineering. The distribution reflects the prevailing view during the period the corpus was assembled. The retrieval layer indexes the enterprise's own records, which contain a certain distribution of prior senior engineering hires, performance reviews, and promotion decisions. Those records reflect the enterprise's prior beliefs about senior engineering. The orchestration layer decomposes the query into sub-tasks, each of which inherits the same framing. The evaluation layer measures output quality against benchmarks that were defined by the same population of experts using the same mental model.

At every layer, the premise is not just preserved but amplified. Each layer reduces the ambiguity that might have allowed the premise to be interrogated. The recruiter's unstated assumption becomes the agent's operating frame becomes the retrieval layer's indexing scheme becomes the orchestration layer's sub-task specification becomes the evaluation layer's quality metric becomes the application layer's recommendation. The recommendation arrives in front of the recruiter wearing the authority of the stack. The recruiter looks at it and sees their own assumptions reflected back at them with numerical confidence. They approve it.

This is the core failure mode. It is not a bug. It is the predictable consequence of a layered architecture in which no layer is specifically responsible for asking whether the premise is sound. The layers were designed for a division of labor that made sense when each layer was a different engineering specialty. They were not designed to accommodate the cognitive function that catches wrong premises, because that function does not have a layer.

Where does the premise-checking function live in human cognition? In humans, it does not live in any single subsystem. It is distributed across many mechanisms: peripheral awareness that registers anomalies, informal conversation that surfaces alternative framings, direct encounter with outcomes that contradicts expectations, the accumulated judgment that experts call taste. These mechanisms operate slowly, informally, and unreliably. They are, nonetheless, the mechanisms by which humans have historically caught wrong premises and corrected the systems that were running on them. None of these mechanisms has an analog in the agent stack. The agent stack was not designed to include them, and it is not clear how they could be added without breaking the abstraction boundaries that make the layered architecture work as engineering.

The question this raises is uncomfortable. If premise-checking is a function the stack cannot perform, and if the queries that fail catastrophically are exactly the queries whose premises are silently wrong, then the stack's failure mode is not a tail risk. It is the modal failure mode. It is what will happen in the general case, not the edge case. The edge case is the query whose premise happens to be right, in which case the stack performs beautifully. The general case is the query whose premise is a little bit wrong, in which case the stack performs beautifully at the wrong task. The difference between the two is invisible to the stack. It is, for a period of time that may be long, invisible to the human at the top of the stack as well, because the human was the one who originally supplied the premise.

The Expert Marketplace Thesis

The most common response to the kind of argument this chapter is making goes something like this. Yes, the stack as currently deployed has the limitation you describe. But the market will produce specialization. Expert systems will emerge for specific domains. Medical agents will be trained by medical experts on medical data. Legal agents will be trained by legal experts on legal data. Financial agents will be trained by financial experts on financial data. The domain expertise will be built in. The premise-checking function will be performed, effectively, by the experts who built the domain-specific system.

This thesis has the advantage of sounding plausible. It aligns with how specialization has worked in earlier technology transitions. It offers enterprises a reassuring story in which the current limitations are transitional and the mature agent economy will have the expert substrate that makes premise-checking automatic. It is the thesis under which a significant fraction of the current investment in applied AI is being made.

The thesis is the expert-shaped version of the same failure mode the chapter has been describing. To see why, we need to ask what an expert-trained agent actually is.

An expert-trained agent is a system whose training data has been curated by domain experts, whose evaluation criteria have been defined by domain experts, whose outputs have been benchmarked against the judgment of domain experts. In effect, the agent encodes the collective judgment of the expert population that built it. When the agent is queried, it produces the output that the expert population would have produced. This is precisely what the vendors and the experts would hope for.

Consider what this means for the premise-checking function. The expert-trained agent inherits the premises of the expert population at the time the training was done. The expert population has its own premises, which are a product of its professional training, its institutional incentives, its recent experiences. Those premises are not neutral. They reflect the current state of the profession, which is the current state of a social system that has its own blind spots, its own conventional wisdom, its own unexamined assumptions about what matters.

When the premises of the expert population happen to correspond to the premises the enterprise actually needs, the expert-trained agent produces excellent recommendations. When the premises of the expert population diverge from what the enterprise actually needs, which happens whenever the environment is shifting faster than the profession can update, the expert-trained agent produces excellent recommendations to the wrong question, with the additional authority conferred by the expert substrate.

The medical profession has certain conventional wisdoms about which conditions are time-critical. The legal profession has certain conventional wisdoms about which arguments carry weight with current courts. The financial profession has certain conventional wisdoms about which risks are worth underwriting. These conventional wisdoms are updated through the ordinary mechanisms of professional discourse, which operate slowly, often with significant lag, and sometimes with systematic resistance to the updates the environment demands. An expert-trained agent encodes the conventional wisdom at a particular moment and deploys it at agent throughput, which is orders of magnitude faster than the profession's own updating mechanism.

The result is not a system that has internalized expert judgment. It is a system that has frozen expert judgment at a point in time and distributed it across a scale at which the profession's informal correction loops, the conversations at conferences, the arguments in journal clubs, the cases that force practitioners to revise their intuitions, cannot reach. The expert marketplace thesis does not solve the premise problem. It relocates the premise to the expert population, where it enjoys the same invisibility it had in the user's head, with the added feature that now the premise is baked into a system that executes at speeds the expert population's own error-correction loops cannot match.

There is a subtler version of the thesis that deserves its own treatment. This version argues that the expert-trained agent will be continuously updated with new data, that the feedback loops will be built in, that the system will learn. In principle, yes. In practice, the learning is constrained by what the system is able to observe about its own outputs and their consequences. The system can observe outputs. It can observe, with some lag, the proximate downstream events that follow from the outputs. It can observe, with more lag and less fidelity, the distal outcomes that the outputs were meant to produce. What the system cannot observe, except through a chain of inference that itself inherits the system's premises, is whether the question it was asked was the question it should have answered. That is the premise problem, and it does not yield to continuous learning unless the learning loop includes a component that is not itself inside the premise. No such component exists in the current architecture.

The Four Cognitive Failures

The premise problem is not one failure mode. It is a family of failure modes, each of which corresponds to a different way in which a premise can be wrong, and each of which the Risk Efficacy framework identifies as a dimension along which cognition can be miscalibrated. In humans, the failure modes appear one at a time, usually, and are addressed by specialized correction mechanisms that evolved over long periods of institutional history. In the agent stack, the failure modes appear simultaneously, compound each other, and are not addressed by any dedicated correction mechanism.

The first failure is miscalibration. The system's expressed confidence does not match its actual reliability. An agent that is eighty percent confident in an output should be right about eighty percent of the time across the population of queries at that confidence level. The agents currently in deployment are not calibrated in this sense, and the miscalibration is systematic: confidence is a function of fluency and the consistency of the generated response with its own context, not a function of actual correspondence to reality. An agent can be supremely confident in a wrong answer when the wrong answer is fluent and internally consistent, which is most of the time. The user, who is not in a position to independently assess correspondence to reality, treats the confidence signal as informative. It is informative about the system's fluency. It is not informative about the system's accuracy. This is the Calibration failure, and it maps, in the historical chapters that follow, onto the case of indulgences and the printing press.

The second failure is failure of informed navigation. The system cannot, except in trivial cases, acknowledge the specific things it does not know and adjust its recommendation to accommodate those unknowns. When an agent produces a recommendation, the recommendation is a synthesis across the available information. The synthesis does not distinguish sharply between what the agent has strong evidence for and what the agent has thin evidence for or no evidence for. The recommendation has a uniform surface quality. The user cannot tell which parts of the recommendation are well-supported and which parts are confabulated filler. The system has not performed the internal accounting that would let it mark its own confidence non-uniformly across the output. This is the Informed Navigation failure, and it maps onto the case of securitization and the collapse of the mortgage-backed securities market in 2008.

The third failure is failure of resilience. The system has no mechanism for noticing when it has made a wrong turn and correcting its course. Once a chain of reasoning has been initiated on a wrong premise, the subsequent steps will be internally consistent with the premise. Each step will look reasonable. The overall chain will look coherent. There is no layer in the stack that compares the trajectory of the reasoning to the target the reasoning was supposed to reach and flags the divergence. In complex agentic workflows, where a primary agent delegates to auxiliary agents that delegate to further auxiliary agents, the wrong premise propagates through a tree of sub-tasks, each of which executes faithfully against its local instructions, producing a deeply internally consistent and deeply wrong result. This is the Resilience failure, and it maps onto Soviet central planning, where the production plan locked in an economic model that could not update against accumulating local shortages.

The fourth failure is failure of outcome achievement. The system produces outputs that look correct at the layer of the output but do not correspond to the outcomes the enterprise deployed the system to produce. A hiring recommendation is evaluated by whether the candidate was hired, not by whether the candidate succeeded in the role. A credit recommendation is evaluated by whether the loan was approved, not by whether the loan performed. A clinical triage recommendation is evaluated by whether the triage decision was implemented, not by whether the patient had a better outcome than they would have had under a different triage decision. The evaluation layer of the stack is measuring proximate outputs because proximate outputs are measurable at the cadence of the agent loop. The distal outcomes, which are what the enterprise actually cares about, are measurable only on longer time horizons and often not cleanly attributable to any single recommendation. This is the Outcome Achievement failure, and it maps onto the harvest reporting chain under the Great Leap Forward, where the reporting system was certifying outputs, the grain figures the cadres submitted, without measuring whether the outputs corresponded to the outcome, the grain that actually existed in the fields.

These four failures are not independent. A miscalibrated system will navigate poorly because it does not know what it does not know. A system that navigates poorly will fail to adapt because its model of the environment was wrong to begin with. A system that cannot adapt will produce outcomes that diverge from intent over time, because reality does not wait for the system's premises to catch up. The failures compound. At scale, they compound quickly, because the throughput at which the failures propagate is high and the error-correcting mechanisms are slow.

What Cannot Be Fixed Inside the Stack

One natural response to the argument so far is to say: very well, these are problems, but they are problems that can be addressed inside the existing architecture. Add calibration layers. Add uncertainty quantification. Add outcome tracking. Add adversarial evaluation. The vendor ecosystem is already building all of these. The criticism is premature. The mature stack will have the missing components.

The chapter will close by taking this response seriously and explaining why the response is only partially right.

The response is right that the vendor ecosystem is building the missing components. Calibration methods are an active research area. Uncertainty quantification is being built into production systems. Outcome tracking is the focus of most of the observability work in enterprise AI. Adversarial evaluation is standard in the frontier labs. These are real improvements, and they will reduce the magnitude of the failure modes the chapter has described.

The response is wrong about where those improvements will leave us. The improvements reduce failure within the space of queries the stack can, in principle, evaluate. They do not address the failure mode that sits outside that space, which is the failure that occurs when the premise of the query is silently wrong. A calibrated system can still be calibrated on the wrong task. An uncertainty-quantified system can still be uncertain about the wrong things. An outcome-tracked system can still be tracking the wrong outcomes. An adversarially evaluated system can still be robust to the wrong adversary. Each improvement raises the performance of the system within its premise. None of the improvements addresses whether the premise is sound.

The premise-soundness function is the function no layer of the stack currently performs. The historical cases the rest of this book examines are cases in which a civilization discovered, late and at cost, that the premise-soundness function in its institutional architecture had been eroded by a change in the medium, and that the new medium had no replacement function built in. In each case, the correction arrived eventually. It arrived as an institutional innovation that sat on top of the amplified medium and performed the premise-soundness function the amplified medium could not perform for itself. It arrived too late to prevent the collapse. It arrived in time to prevent the next collapse, until the next medium arrived and stripped it out again.

The agent economy is the current instance. The premise-soundness function has not yet been built. The historical pattern is that the function gets built after the first several collapses, calibrated to the specific shape of the damage that the collapses produced. The pattern does not have to hold this time. It would require a civilization that has learned to recognize the pattern while it is still in the amplification phase and to build the correction layer before the collapse. Whether that civilization is the one we are currently living in is an open question. The rest of this book is an attempt to make the question answerable by showing the reader the pattern clearly enough that they can see their own moment inside it.

What follows is the diagnostic framework in its generalized form. It was developed for the measurement of individual decision-making. It applies equally to institutional decision-making and to infrastructural decision-making, which is the extension the rest of the argument depends on.

Chapter 2: The Four Signatures

A Framework That Travels

The diagnostic vocabulary this chapter introduces was developed to solve a measurement problem at the individual level. Organizations could not reliably identify the people in their populations who were making good decisions under uncertainty, because the proxies they relied on, outcomes, credentials, and confidence, were systematically decoupled from decision quality itself. The Risk Efficacy framework offered an alternative. Measure the four dimensions along which any decision process can be observed and assessed, Calibration, Informed Navigation, Resilience, and Outcome Achievement, and a picture of decision quality emerges that the conventional proxies obscure.

The framework was first developed for individuals, then extended to populations whose decision-making the standard measurement systems systematically misread, an extension developed in *The Neurodivergent Advantage* (Wu, 2026). That extension demonstrated something the original presentation had only gestured at. The framework is not a property of human psychology. It is a property of decision processes, and decision processes can be instantiated in substrates that are not human at all.

This chapter develops the implication. The four dimensions that apply to a human decision-maker also apply to an institutional decision-making system and to a computational one. What changes across the three scales is not the framework itself. What changes is the mechanism by which each dimension is implemented, the mechanism by which each dimension can fail, and the mechanism by which each dimension can be corrected. A miscalibrated individual, a miscalibrated institution, and a miscalibrated agent stack are instances of the same phenomenon with different carriers. Without a generalized framework, the historical chapters that follow and the agent-economy chapters that close the book would read as a collection of unrelated stories about unrelated failures. With it, they read as instances of a single architectural pattern.

What follows is the framework in its generalized form. The treatment is condensed, since the argument ahead does not require re-deriving it from first principles, and expanded in one respect. Each dimension is presented at three scales, individual, institutional, and infrastructural, with a brief account of how it operates and how it fails at each.

Calibration

Calibration is the relationship between expressed confidence and actual accuracy. A calibrated decision-maker assigns probability seventy percent to claims that turn out to be true seventy percent of the time, probability ninety percent to claims that turn out to be true ninety percent of the time, probability ten percent to claims that turn out to be true ten percent of the time. Calibration is the property that lets other agents treat the decision-maker's confidence signals as informative. When calibration is preserved, a confidence signal conveys information about the likelihood of an outcome. When calibration is lost, the signal conveys information about the speaker, or about the speaker's relationship to the social environment in which the speaking occurs, or about nothing in particular. The signal degrades from an instrument of coordination into noise that may or may not carry useful content.

At the individual level, calibration is a property that can be measured and trained. Forecasting research has established that calibration is separable from other cognitive abilities. It has also established that people can improve their calibration substantially with feedback, if the feedback is timely, specific, and connected to the predictions that generated it. Expert forecasters in domains like weather and sports betting show near-perfect calibration because they operate in environments that provide exactly the feedback structure that calibration training requires. Experts in domains without that feedback structure, like long-range geopolitical forecasting or venture investing, show calibration that is often worse than a naive baseline, because the environment never told them they were miscalibrated in a form they could respond to.

The individual-level calibration failure is familiar. A person expresses high confidence about a claim. The claim turns out to be wrong. Other people, who took the expressed confidence as informative about the claim, find themselves having acted on bad information. The social cost of expressing high confidence in wrong claims, over time, disciplines most individuals into moderate calibration. They learn to say "I am not sure" when they are not sure, because the people around them have learned to treat their certainty signals skeptically when those signals have a history of being wrong. The informal social feedback loop is the mechanism by which individual calibration is maintained across the population.

At the institutional level, calibration is the property that lets the institution speak with one voice on questions that require coordination. The institution expresses a view. Other institutions act on the view. When the view turns out to be right at the rate the institution's confidence signaled, the system works. When the view turns out to be wrong more often than the confidence signaled, other institutions stop treating the confidence as informative. Institutional calibration is maintained, when it is maintained, by internal mechanisms that surface dissent, by external accountability structures that impose costs on persistent miscalibration, and by the professional norms that discipline the individuals who speak on behalf of the institution.

The institutional-level calibration failure is the one that generates most of the policy crises we recognize. An intelligence agency expresses high confidence in a claim about weapons of mass destruction. A central bank expresses high confidence in a claim about inflation being transitory. A public health authority expresses high confidence in a claim about transmission dynamics. The confidence signals get treated as informative by the institutions downstream of the original claim. Policy is made. Resources are allocated. Lives are reorganized. And then the claim turns out to have been confidently wrong. The institution that produced the wrong claim suffers a loss of credibility. The institutions that acted on the claim suffer consequences proportional to their action. The cost of institutional miscalibration is larger than the cost of individual miscalibration because more action is coordinated on the basis of institutional signals.

At the infrastructural level, calibration becomes a property of the communication medium itself. The medium carries confidence signals from producers to consumers. The medium has a set of features that, in historical cases, have generally worked to discipline the confidence signals it carries: the physical proximity that made the producer identifiable to the consumer, the feedback loops through which repeated interactions registered the producer's track record, the institutional filters through which claims were vetted before propagation. When a medium changes, these features may or may not be preserved. If they are not preserved, the medium carries confidence signals without the disciplinary apparatus that kept them informative. The signals propagate faster than the correction loops can catch up. Consumers act on signals whose informational content has decoupled from what the signals purport to convey. At sufficient scale, the population of signals becomes epistemic noise.

The printing press is the historical case that illustrates the calibration failure most cleanly. The medium carried confidence signals at a scale and speed the prior correction mechanisms had been calibrated to, and it produced an infrastructural miscalibration that took approximately a century to absorb. The indulgence certificates, which were the specific confidence signals in question, were well-calibrated in the local pre-print economy and catastrophically miscalibrated in the continental print economy. The calibration did not fail because the individuals producing the certificates had changed their character. It failed because the correction loop that had disciplined the production of the certificates operated at the throughput of the pre-print medium and could not operate at the throughput of the new one.

In the agent stack, the calibration failure lives at the compute and inference layer in a particular sense, and at all other layers in a different sense. The compute and inference layer takes any query and produces an output whose confidence signal is a function of the model's internal state. The confidence signal is produced at high throughput. The feedback loop that might discipline the confidence signal, the loop in which the output is compared against reality and the system is updated, operates, if at all, at a throughput many orders of magnitude lower than the confidence signals it is supposed to discipline. The calibration of the signal is a property of the training distribution, and the training distribution is an empirical fact about the past, not a real-time correspondence with the present. The calibration can therefore be arbitrarily bad, and the badness is not visible inside the stack because the stack has no mechanism for comparing its signals against ground truth at the speed the signals are produced.

Informed Navigation

Informed Navigation is the process by which a decision-maker gathers information, weighs evidence, and explicitly accounts for what is not known. It is the dimension along which people with careful methodological practice differ from people with intuitive confidence. The distinction is not between caution and boldness. It is between decision processes that explicitly enumerate their unknowns and those that fold unknown into known without noticing. The former produce decisions that can be updated when new information arrives, because the unknowns were flagged. The latter produce decisions that cannot be updated, because the unknowns were absorbed into the framing.

At the individual level, Informed Navigation is the skill of the decision-maker who can distinguish signal from noise, who can identify the information they need but do not have, who can quantify the residual uncertainty after the available information has been integrated. This skill is not the same as intelligence. It is closer to a discipline, maintained by habit and by the meta-cognitive work of asking, at each stage of a decision, what the gaps are and whether the gaps are material. The individuals who score high on Informed Navigation tend to be slow. They tend to produce decisions that are harder to communicate because the decisions carry the explicit acknowledgment of what the decision-maker does not know. They tend to be undervalued in environments that reward speed and certainty. They tend to be the people whose calls, in retrospect, turned out to anticipate the failures that the fast and certain decision-makers missed.

The individual-level Informed Navigation failure looks like this. A person confronts a decision. They have some information. They have awareness of certain gaps. They have awareness of other gaps. They have no awareness of a third category of gaps. They integrate the information, the first category of gaps, and an approximation for the second category of gaps into a recommendation. The recommendation does not reflect the third category of gaps, because the third category was not visible to the decision-maker's model. The decision gets made. The decision's performance then depends, in part, on whether the third category of gaps matters for this particular decision. Sometimes it does not, and the decision is fine. Sometimes it does, and the decision produces an outcome the decision-maker finds baffling, because nothing in the decision process anticipated the source of the failure.

At the institutional level, Informed Navigation is the property of the decision-making system that determines whether the institution's recommendations reflect what the institution knows and, just as importantly, what the institution does not know. Institutions have enormous stores of information. They have analytical capacity. They have the ability to commission research, run experiments, solicit external review. They also have internal mechanisms that reduce decision problems to their known components and discard the unknown components as noise, because acting on unknowns is harder than acting on knowns and the institution's incentive structure often rewards visible action over invisible prudence.

The institutional-level Informed Navigation failure is the one that produces the post hoc investigations that find the institution had the information, the information was on someone's desk, the information would have changed the decision, and yet the decision was made as if the information did not exist. The question is not why the information was not available. The question is why the decision process did not ingest the information in a form that would have changed the recommendation. The answer, usually, is that the decision process was built around a certain set of inputs and the information in question did not fit the input template. It was unstructured data. It was from the wrong department. It was flagged by someone without the authority to require its inclusion. The decision process has an implicit ontology of what counts as information, and anything outside the ontology is invisible, regardless of how relevant it is to the decision.

At the infrastructural level, Informed Navigation becomes a property of how information travels through the system that aggregates it. Information originates somewhere, typically close to the ground, in the observations of people doing the actual work. It travels, through some chain of intermediaries, to the point at which decisions that depend on it are made. Along the way, it is summarized, aggregated, translated, abstracted. Each transformation is, in principle, an opportunity to preserve relevant information and discard noise. Each transformation is, in practice, also an opportunity to lose information that does not fit the receiver's template, to smooth out anomalies that the receiver does not know how to interpret, to encode the sender's prior beliefs about what matters into a summary that the receiver will treat as a neutral description.

The securitization of residential mortgages in the decade before 2008 is the case that illustrates this failure most clearly. A mortgage is originated by a loan officer who knows something about the borrower, the neighborhood, the local economy, the specific features of the loan. The mortgage is packaged into a security, which abstracts the individual loan into a statistical feature of a pool. The security is rated by an agency, which abstracts the pool into a credit rating derived from a generic model. The rated security is sold to an investor, who owns a tranche of the pool without any direct knowledge of any of the underlying loans. At each step, the navigation function, the asking of what is not known about this specific asset, has been handed off. At the final step, the investor is navigating a generic model, not a set of loans. The investor's Informed Navigation is over the space of tranches and correlations. The loan officer's Informed Navigation is over the space of borrowers. The two spaces are connected only by the assumption that the statistical features of the pool capture the relevant information about the underlying loans. When that assumption is wrong, the navigation function has no mechanism for detecting the error, because the error lives in the information that was abstracted away at the transformation steps.

In the agent stack, the Informed Navigation failure lives at the orchestration and multi-agent workflow layer, in a particular sense. When a query is decomposed into sub-tasks and the sub-tasks are delegated to auxiliary agents, each auxiliary agent receives a specification that assumes the premise. The auxiliary agent executes against the specification. It has no mechanism for flagging that its sub-task would produce a different answer if the premise were different, because it has no awareness of the premise. The orchestration layer composes the sub-task outputs into a recommendation. The recommendation inherits the composition logic's assumption that if the sub-tasks are each correctly executed, their composition is correctly executed. This assumption is false in the specific case where the composition is being done in service of a wrongly-framed master query. The Informed Navigation function, the explicit enumeration of what is not known about the relationship between the master query and the sub-tasks, does not exist anywhere in the stack.

Resilience

Resilience is the property that lets a decision-maker absorb unexpected outcomes without catastrophic failure. It is the dimension along which decision systems differ in their capacity to update, to adapt, to recover from errors. A resilient system, having discovered that a decision did not produce the expected result, uses the discovery to revise its model and its behavior. A non-resilient system, having discovered the same, either does not discover it, or discovers it and fails to update, or updates superficially without reaching the underlying model that produced the bad decision.

At the individual level, Resilience is the capacity of a person to take information that disconfirms their prior beliefs and integrate it into a revised understanding, rather than dismissing the information or explaining it away. The literature calls this belief updating. The practical version is less formal. A resilient decision-maker, when something goes wrong, asks what they were wrong about, how they were wrong, and what the error reveals about their prior mental model. They treat errors as data. They revise accordingly. A non-resilient decision-maker treats errors as noise, attributes them to bad luck, or finds a way to rationalize the error as consistent with the prior model. The non-resilient decision-maker accumulates errors without learning from them, which means the next decision inherits all the flaws of the prior one plus whatever new flaws the new situation introduces.

The individual-level Resilience failure is the pattern that Kahneman and his colleagues documented extensively. A person makes a decision under uncertainty. The decision produces an outcome. The person's attribution of the outcome depends on whether the outcome was good or bad, not on the quality of the decision process that produced it. Good outcomes are attributed to skill. Bad outcomes are attributed to luck. The attribution pattern means that the person learns almost nothing from their own track record, because the signal they would need, which decisions were good and which were bad, has been systematically distorted by the attribution bias. The decision-maker can accumulate years of experience without the experience actually improving their decision-making, because the experience was ingested through a filter that preserved the illusion of competence.

At the institutional level, Resilience is the property of the organizational system that determines whether the organization can learn from its failures. Institutions have formal mechanisms for learning, the after-action review, the postmortem, the lessons-learned memo. They also have informal mechanisms that undermine the formal ones: reputational incentives that make truth-telling about failures costly, political dynamics that make assigning responsibility contentious, institutional memory that decays faster than the time horizons on which institutional decisions play out. The result is that institutions often know they failed but do not know why, or know why but cannot act on the knowledge, or act on the knowledge only by changing the cosmetic features of the decision process without changing the underlying model that produced the failure.

The institutional-level Resilience failure is the pattern by which the same institution produces the same category of failure repeatedly across decades, despite extensive documentation of each prior failure. The intelligence community's repeated failures in analytical forecasting. The financial regulatory apparatus's repeated failures in recognizing systemic risk. The public health establishment's repeated failures in pandemic response. Each failure is followed by an investigation, which identifies specific factors, which are addressed through specific reforms, which do not prevent the next failure from exhibiting the same structural features. The structural features are not visible to the institutional review, because the structural features are the institution's own operating model, and the institution's review is conducted from inside that model.

At the infrastructural level, Resilience becomes a property of the system that determines whether the model can update against feedback. The infrastructural model is the shared representation of reality on which multiple decision-makers operate. The model is updated, when it is updated, through some mechanism that takes signals from the environment and revises the model accordingly. If the mechanism has high bandwidth, the model stays close to reality. If the mechanism has low bandwidth, the model drifts from reality over time, and the decisions made on the model become progressively worse without the decision-makers noticing, because they are operating inside the model rather than outside it.

Soviet central planning illustrates this failure at full scale. The Soviet planning apparatus constructed a model of the economy. The model was updated through the reports that traveled from enterprises to ministries to the central plan. The reports were produced by people whose incentives were to confirm the prior plan rather than to surface its errors. The model therefore updated in a direction that tracked the planners' prior beliefs, not the actual economic reality. Over decades, the model and reality diverged. Production targets became increasingly detached from what the economy could produce. Allocations became increasingly detached from what the population needed. The system continued to operate, because the system's operation did not depend on the model's correspondence to reality. It depended only on the model's internal consistency, which was maintained with considerable effort. The failure was not visible to the system. It was visible only to the population that lived inside the system and that had no mechanism for making its observations register in the model.

In the agent stack, the Resilience failure lives at the memory and context layer. The memory layer encodes what the system has learned from prior interactions, what the system knows about the enterprise, what the system treats as the current state of the world. The encoding is not a direct correspondence with reality. It is a compressed representation of reality, filtered through the interactions that produced it, shaped by the retrieval logic that governs which memories are surfaced and when. When reality shifts, the memory does not automatically shift with it. The memory retains the pattern it encoded, and the pattern shapes every subsequent retrieval. The agent operating on the memory is operating on a model that has a certain correspondence with past reality and a progressively worse correspondence with current reality. The system has no mechanism for detecting the divergence, because the divergence is defined in terms of a reality the system does not independently observe.

Outcome Achievement

Outcome Achievement is the final dimension of the framework. It is the relationship between the outputs a decision process produces and the outcomes the process was instituted to achieve. A well-converting system produces outputs that, on average, lead to the intended outcomes. A poorly-converting system produces outputs that look appropriate on their face but fail to translate into the outcomes the system is ostensibly serving. The distinction between outputs and outcomes is the distinction between what the system does and what the system is for. Many systems, across scales, can continue to produce outputs long after the outputs have ceased to convert into outcomes, because output-production can be self-sustaining whereas outcome-production requires continuous contact with the reality the outcomes live in.

At the individual level, Outcome Achievement is the property of the decision-maker's performance that determines whether their decisions, across time, produce the results they were meant to produce. The individual decision-maker operates on a model of what their decisions are supposed to accomplish. The model can be more or less aligned with the actual purpose of the decision-making. A trader who optimizes for the P&L over a given quarter may be converting outputs into outcomes, if the P&L is a fair measure of what the trading function is supposed to produce, or may be converting outputs into a proxy that diverges from outcome, if the P&L can be inflated by taking risk that is not captured in the measure. The individual's Outcome Achievement depends on the gap between what the individual is measured on and what the organization actually needs from the individual's role. When the gap is small, the individual's outputs are outcomes. When the gap is large, the individual's outputs are something else, and the something else can be cultivated for years without the underlying outcome being served.

At the institutional level, Outcome Achievement is the property of the institutional performance system that determines whether the institution's outputs correspond to the outcomes the institution is supposed to be producing. Institutions are graded on outputs because outputs are what can be measured. Financial institutions are graded on returns. Hospitals are graded on procedure volumes and complication rates. Universities are graded on admission selectivity and graduate placement. Regulators are graded on enforcement actions. The grading system does not measure the underlying outcomes, investment capital allocated to productive uses, patients made healthier across their life courses, students prepared for lives of consequence, markets protected from systemic failure, because those outcomes are harder to measure and longer in coming. The institution optimizes for the outputs it is graded on. The outputs diverge from the outcomes, slowly, over time. At some point the divergence becomes visible, either because the outcomes become catastrophically bad or because the divergence itself becomes politically salient. The institution is then grilled over the divergence, and a round of reform produces a new set of output measures that are, for a time, closer to the outcomes, until the measures become the target and the divergence restarts.

At the infrastructural level, Outcome Achievement becomes a property of the reporting chain itself, which is the mechanism by which outputs from one layer are converted into inputs for the next layer. The reporting chain is supposed to transmit information about outcomes. In practice, the reporting chain transmits information about whatever outputs the previous layer was able or willing to produce. If the outputs correspond to outcomes, the chain works. If the outputs are disconnected from outcomes, the chain is transmitting a representation of reality that is increasingly divorced from reality itself, and every layer further up the chain is operating on inputs that have been systematically distorted by the chain's own structure.

Harvest reporting under the Great Leap Forward illustrates this failure at its most extreme. The village cadre reported grain production to the county. The county aggregated the reports and sent them to the province. The province aggregated the reports and sent them to Beijing. At each level, the report was an output: a number representing the purported grain production of the unit below. The number was not an outcome: it did not correspond to the grain that actually existed in the fields. The divergence grew at each level, because each level had incentives to certify upward compliance with the production targets, and each level's output was the input to the next level's aggregation. At the top of the chain, Beijing had a national grain figure that was a fiction, constructed by the summation of fictions. The planning decisions made on that figure were outputs of a decision process whose inputs bore no correspondence to the outcomes the decisions were ostensibly about. Tens of millions of people starved in a famine the reporting chain could not see.

In the agent stack, the Outcome Achievement failure lives at the evaluation and observability layer. The evaluation layer measures the system's outputs: the recommendations, the completions, the code, the summaries. The measurement is performed against a set of metrics that were chosen for their measurability and their legibility to human reviewers. The metrics are proxies for the outcomes the enterprise deployed the system to produce. The proxies were chosen at a moment when the enterprise had a certain model of what the system was for. The model has a shelf life. The proxies drift from the outcomes at a rate determined by how fast the environment changes. The evaluation layer continues to certify that the system is performing well, because the system is performing well against the proxies. The outcomes, meanwhile, are diverging from what the proxies capture. The enterprise does not know this, because the enterprise is looking at the dashboard the evaluation layer produces, and the dashboard shows the proxies, not the outcomes. By the time the divergence becomes visible, either because the outcomes become catastrophically bad or because a competitor produces a system whose outcomes are visibly better, the enterprise has been operating on a fiction for months or years.

The Framework as a Diagnostic

The four dimensions are independent in principle. A system can fail on Calibration without failing on Navigation, or fail on Resilience without failing on Achievement. In practice, the dimensions tend to correlate: a system that is miscalibrated will navigate poorly, a system that navigates poorly will fail to adapt, a system that fails to adapt will produce outputs that diverge from outcomes. The correlations are not accidental. They reflect a deeper fact about decision systems, which is that the function of each dimension depends on the functions of the others. A calibration signal is only useful if it is embedded in a navigation process that uses it. A navigation process is only useful if it can be updated through resilience mechanisms. A resilience mechanism is only useful if it is connected to the outcomes the system is supposed to produce.

The four dimensions are, in this sense, the four ways a cognitive system can be well-calibrated to reality, and the four ways it can become decoupled from reality. The historical chapters that follow each isolate one of the four dimensions and show how its failure, amplified by a specific medium, produced a specific collapse. The isolation is analytical, not empirical. In each case, the actual collapse involved failures on multiple dimensions. But in each case, one dimension was the initiating failure, the one whose degradation opened the door for the others, the one whose restoration would have been the point of intervention most likely to have contained the collapse.

Calibration, the printing press, indulgences.

Informed Navigation, securitization, the 2008 financial crisis.

Resilience, central planning, the collapse of the Soviet economy.

Outcome Achievement, the reporting chain, the Great Leap Forward famine.

Four pillars, four cases, four layers of the agent stack. The symmetry is intentional. It is also, as I will argue in the concluding parts of the book, empirical. The layered architecture of the agent stack maps onto the four pillars not because the designers of the stack had the pillars in mind, but because the stack was designed to solve the same cognitive problems the pillars describe, and each layer is solving its piece of the problem in isolation. The isolation is the source of the failure. The pillars, applied across layers, are the vocabulary for making the failure visible.

With the framework in hand, we turn to the historical record. The first case is the case that shaped much of the early modern world. It began with a practice that was old, local, and cognitively manageable. It became, through the introduction of a new medium, a continent-wide failure that took a century to absorb and that left behind the institutional infrastructure on which the modern state was built.

Part II: The Pattern Has Happened Before

Chapter 3: Indulgences and the Press

The Local Economy of Forgiveness

In the decades before Gutenberg, an indulgence was a small transaction in a small economy. A villager who had accumulated some sense of spiritual debt, whether through sins confessed or sins weighing silently on conscience, could obtain from the local clergy a certificate that was said to reduce or remit the temporal punishment owed for those sins. The transaction required a payment, a local priest, and a certificate that was often handwritten, stamped with the authorizing bishop's seal, and presented to the buyer in a ceremony that made the transaction feel appropriate to its stated purpose.

The practice was old by the fifteenth century. It had evolved across centuries of theological elaboration. It was, in its original conception, a mechanism by which the Church could channel penitential energy into specific works of piety, support for the poor, funding for crusades, construction of cathedrals, and in exchange could offer the penitent some assurance that the penance had been credited to their spiritual account. The theology was intricate, and the theological defenders of the practice were at pains to distinguish what an indulgence did, which was modify the temporal penalty for sin, from what it did not do, which was absolve the sin itself. The distinction mattered to theologians. It often mattered less to buyers.

The economy in which the practice operated had a specific set of features that, taken together, kept the calibration between what the certificates promised and what buyers believed they purchased within a tolerable range. The priest who issued the certificate was usually known to the buyer. The buyer was usually known to the community. The transaction took place within a web of relationships that extended before the purchase and would continue after it. If a priest was understood to be selling certificates whose theological claims outran what the Church actually authorized, the community would know, because the community knew the priest, and the community would discipline the priest through the slow, informal mechanisms by which small communities discipline their members. The bishop who supervised the priest would hear complaints. The visiting papal legate would be told. The reputation of the priest, and through the priest the reputation of the diocese, would register consequences for miscalibration.

None of this made the system rigorous in any modern sense. The informal correction loop was slow, uneven, and subject to capture by the same local power structures that everything else in the medieval world was subject to. But it was a correction loop, and it had the property that matters for this chapter: it could catch a miscalibration before the miscalibration scaled beyond the local community. A priest could not sell five thousand indulgences to villagers in a neighboring diocese without someone noticing, because the priest could not physically travel to that many villages, and the villagers in the next diocese had their own priest, whose certificates they would buy if they wanted to buy any. The system was locally bounded, and the local boundaries were the feature that made the calibration sustainable across centuries.

What I want the reader to hold in mind, as we move into the story of how this local system became something else, is that the original architecture was not innocent. Individual priests were sometimes corrupt. Individual bishops sometimes profited from the practice in ways that embarrassed the Church. Individual theologians sometimes voiced concerns about the slippage between what the certificates technically promised and what buyers believed they received. These local failures are precisely the failures the community's informal correction loops were designed to catch. The loops caught some and missed others. The result was a messy, approximate calibration that held up well enough for the practice to survive across centuries of theological and institutional change.

The point of the story is not that the old system was good. The point is that the old system was locally bounded, and the local boundaries contained failures that would otherwise have compounded across the larger ecclesiastical body. The next part of the story is what happened when a new medium stripped the local boundaries out.

The Amplifier

The first printed indulgence certificates appear in the historical record in the 1450s, within a decade of Gutenberg's initial work. Their presence in the early output of the German print industry is not a curiosity. It reflects the economic reality of early printing, which needed high-volume, low-complexity jobs to amortize the cost of type-setting and paper procurement, and found them in the standardized certificates the Church had been producing by hand for centuries. The printer who had capacity between more demanding jobs could run off hundreds or thousands of identical certificates, each one visually consistent, each one professionally produced, each one blank in the places where a buyer's name and a priest's signature would later be inserted.

The effect was not immediately scandalous. The initial use of printed indulgences was, in intent, a mere efficiency improvement. The theological claims on the certificates were the same as those on the handwritten ones. The issuing bishops were the same. The authorizing seals were the same. The only change was that the production process was faster and the certificates were identical, which, given the

Church's interest in consistency, was considered a feature rather than a concern.

What the adopters of printing did not initially appreciate, and what we can see with the clarity of hindsight, is that the printing press did more than speed up production. It changed the relationship between the production of certificates and the correction mechanisms that had disciplined the handwritten economy. Three changes, specifically.

First, the printed certificate was detached from the priest who issued it. The handwritten certificate had carried, in its physical artifact, a trace of the specific priest who had prepared it. The handwriting was that priest's handwriting. The wording, within theological constraints, reflected that priest's phrasing. The physical object was an index of the human being who had produced it. The printed certificate had no such index. It was typographically identical to every other certificate in the print run. The specific priest who later filled in the buyer's name was, in the eye of a buyer who had never met the priest, interchangeable with any other priest.

Second, the printed certificate was portable across geographies at a scale the handwritten system had not accommodated. A boxed run of five thousand certificates could be transported to a distant diocese and deployed there by pardoners, the itinerant sellers who specialized in marketing indulgences to populations without existing relationships to the sellers. The pardoners were not the local priest. They were salesmen. They had incentives to maximize the number of certificates sold. They did not have the long-term relationship with the buying community that had disciplined the handwritten economy.

Third, and most consequentially, the printed certificate increased the total volume of certificates in circulation by orders of magnitude, which meant that the statements made on the certificates, or by the pardoners who sold them, reached populations far larger than the population any individual priest could have reached in the pre-print economy. The claims that a printed certificate could make, and that a pardoner could amplify, were no longer statements made to one's neighbors. They were statements made to crowds of strangers who had no basis for evaluating the speaker.

The combination of these three changes meant that the informal correction loop, which had kept the pre-print indulgence economy in rough calibration, was no longer functional. The loop had depended on the local community knowing the seller, knowing the other buyers, and having slow, repeated opportunities to observe whether the seller's claims matched what the seller actually delivered. The printed economy had none of these features. The seller was a stranger. The other buyers were strangers. The opportunity to observe was a single transaction in a single square on a single market day, after which the pardoner was gone.

Into this calibration gap came the rhetoric that gave the practice the reputation it has kept. Johann Tetzel, the Dominican preacher whose sales campaign in the territories bordering Saxony provoked Luther's response, was not a unique villain. He was the most visible practitioner of what had, by the 1510s, become an ordinary feature of the ecclesiastical economy. Tetzel's campaign featured what historians later called, in his own phrasing or something close to it, the claim that a coin in the collection box produced the soul's release from purgatory at the moment the coin rang against the bottom. Whether Tetzel himself used this phrasing is disputed. That someone in his operation did, and that the phrasing matched a widespread popular understanding of what the printed certificates promised, is not disputed.

The rhetoric mattered because it was the measurable point at which the buyers' understanding of what they were purchasing had decoupled from what the theology actually authorized. Indulgences, in formal Catholic doctrine, remitted the temporal penalty for confessed and repented sin. They did not guarantee the soul's release from purgatory. They did not absolve unconfessed sin. They did not operate as magical transactions independent of the buyer's spiritual state. The printed certificates, read in the way the pardoners' rhetoric encouraged buyers to read them, promised something closer to the magical transaction than to the formal theology. The confidence signal the certificates carried had outrun the reality they were supposed to represent. The Calibration failure, to use the framework's vocabulary, was fully in place by the time Luther's ninety-five theses were posted.

The Correction That Was Stripped Out

The correction mechanism the printing press had disabled was the village-scale relationship between clergy and community. It is worth pausing on the specific features of that mechanism, because they are the features that any calibration-preserving system needs, and they are the features that the agent stack is currently missing in an analogous way.

The village correction mechanism had, first, proximity. The priest who sold the indulgence was physically near the buyer, day after day, year after year. The proximity meant that the buyer could observe the priest's behavior, the priest's other claims, the priest's track record. The confidence signal the priest issued in the indulgence transaction was one data point in a long series of data points the buyer could weight together. If the priest's other claims were unreliable, the buyer had information to discount the current claim. The agent stack has no analog. The user interacting with the agent has no long-term relationship with the specific instance of the system they are using. They have, at best, a relationship with the vendor, which is a relationship with a marketing artifact rather than with the specific computation that produced the current output.

The mechanism had, second, repeated interaction. The priest and the buyer interacted across the rituals of the year: masses, baptisms, weddings, funerals, feast days, confessions. Each interaction was an opportunity for the buyer to update their estimate of the priest's reliability on specific matters. The updating was unconscious and informal, but it was updating. The agent stack does have something like repeated interaction, in the sense that enterprises interact with their agents across many queries. But the repeated interaction does not aggregate into a reliability estimate in the user's mind, because the user's experience of the agent is as a single interface rather than as a series of encounters with a distinguishable agent. The agent does not have a track record the user holds in memory. The agent has a benchmark score the user might have been told once and then forgotten.

The mechanism had, third, community aggregation of reputation. A priest whose behavior diverged from the community's expectations was the subject of conversation. The conversation reached the priest eventually. It reached the priest's superiors. The reputation was a shared resource of the community, maintained through the ordinary social metabolism of a small population living in proximity. The agent stack has no community in this sense. The users of the agent do not know each other. They do not share their experiences systematically. They do not aggregate their observations into a reputation that can discipline the agent's behavior. Each user's experience is private, and the privacy of the experience means that the miscalibrations the user notices get logged, if at all, to a support queue that the vendor reads through a filter shaped by the vendor's interests.

The mechanism had, fourth and most importantly, a feedback loop that operated at the timescale of the claims being made. The claims the pre-print priest made were claims about the spiritual state of the buyer. The feedback on those claims was, at one level, unobservable, because the state of the buyer's soul was not available to the community for inspection. At another level, the feedback was observable through the priest's overall behavior, the priest's other claims, the priest's track record on claims that were empirically checkable. The feedback loop operated at the timescale of human social observation, which was compatible with the timescale at which the priest was making claims. The agent stack operates at a timescale, the millisecond, at which no human feedback loop can function. The feedback that does exist operates on a timescale, the quarter or the fiscal year, at which a vast number of miscalibrated outputs have already been produced and propagated.

What the printing press did, then, was not to introduce the indulgence practice. The practice had existed for centuries. What the printing press did was to multiply the volume of indulgence claims by several orders of magnitude and to distribute those claims across a geography that exceeded the range of the pre-print correction mechanisms by a similar factor. The practice, scaled to the new medium, exceeded the correction capacity of the system that had previously held it in calibration. The Calibration failure was the predictable consequence of the medium-correction mismatch.

The Church's leadership did not see the mismatch in these terms. They saw a theological dispute, a rhetorical excess, a set of pastoral problems that could be addressed through doctrinal clarification and administrative discipline. Each of these diagnoses was accurate at the level it addressed. None of them reached the architectural issue, which was that the medium had outgrown the correction system and that no amount of doctrinal clarification was going to reach into the German countryside fast enough to catch up with the printed certificates already in circulation and the preachers already interpreting them.

The Collapse

The collapse, when it came, did not arrive in the form the Church's leadership was prepared for. Luther's initial intervention in 1517 was not a rupture. It was a request for academic disputation. The ninety-five theses were written in Latin, intended for a theological audience, framed within the conventions of scholarly dispute. Luther himself, through the late 1510s, understood himself to be raising questions within the Church, not launching a schism.

What converted the academic intervention into a continental rupture was the medium that had produced the crisis in the first place. The ninety-five theses were translated into German within weeks. They were printed. They circulated through the German territories in quantities that dwarfed any prior theological pamphlet. By 1520, Luther's major treatises, "Address to the Christian Nobility," "The

Babylonian Captivity of the Church," "The Freedom of a Christian," were being printed in multiple editions and distributed across Europe. The printing press, which had amplified the theologically fragile practice of indulgence-selling into a continental calibration crisis, now amplified the critique of that practice into a continental theological crisis. The same medium that produced the disease produced the symptoms at a scale the patient could not absorb.

The Church responded slowly, inconsistently, and with a range of institutional reflexes that had been calibrated to the pre-print world. It issued papal bulls. It demanded recantation. It eventually excommunicated Luther. Each of these moves made sense in a world where the Church's authority was mediated through the physical infrastructure of the ecclesiastical hierarchy. In the print world, the moves landed on a population that was increasingly able to read arguments for and against the Church's positions in its own language, at its own pace, in formats that did not require the mediation of the clergy. The authority that had depended on the hierarchy's monopoly over the production and distribution of theological arguments now had to compete, in a market it had not previously had to compete in, with producers who used the same medium to offer alternative arguments.

The Reformation, understood in its cleanest institutional terms, was the consequence of this competition. The reformers did not defeat the Church in theological argument. They outcompeted the Church in a medium the Church did not know how to operate in. The confessional map of Europe by the late sixteenth century was largely the map of which territories had favorable conditions for reformist publishing and which did not. Printing shops, university connections, urban literacy rates, the disposition of local sovereigns: these features, more than the intrinsic merit of the theological arguments, determined where the reformed churches took hold.

The casualty count of the consequent wars is a matter for historians to reconstruct and debate. The conventional estimates for the Thirty Years' War alone run into the millions, with some regions losing a third to a half of their population through combat, famine, and disease. These figures do not include the earlier wars of religion in France, the Dutch revolt, the English civil war, the many smaller conflicts that attended the religious settlement of specific principalities. The total human cost of the calibration crisis that began with the printed indulgence was, by any serious accounting, enormous. It was also, from the perspective of the people who lived through it, almost entirely unintended. The initial participants in the printed indulgence economy had not set out to produce a continental religious war. They had set out to exploit an efficiency in the production of devotional certificates.

The Church's eventual response to the crisis, culminating in the Council of Trent in the mid-sixteenth century, addressed the theological and administrative issues the original practice had produced. Trent clarified the doctrine of indulgences, suppressed the most abusive forms of the sale, and instituted reforms of the clerical orders that had made the sale possible. These were meaningful reforms. They were also, as corrections go, a generation late. They arrived after the map of Europe had been redrawn.

The Institution That Got Built

The institution that got built after the collapse was not the Church reformed, though the Church was reformed. It was the modern state. The specific mechanism by which the calibration crisis produced the modern state is worth tracing, because it is the pattern by which every amplifier collapse produces an institutional innovation that sits on top of the amplified medium and performs a correction function the medium could not perform for itself.

The immediate institutional consequence of the religious wars was a settlement, the Peace of Westphalia in 1648, that established the sovereignty of the territorial ruler over the religious question in each territory. The settlement was, in its immediate terms, a compromise: the Catholic and Protestant powers agreed to a durable division of religious allegiance along territorial lines. The deeper consequence of the settlement was the elevation of the territorial state as the unit of political authority, in preference to the transnational religious hierarchy that had previously shared authority with the territorial rulers. The state acquired functions, responsibility for religious peace, responsibility for the regulation of publishing, responsibility for the adjudication of disputes that crossed confessional lines, that had previously been distributed across the Church and the princes.

Over the following century, these functions accumulated into the characteristic apparatus of the early modern state: the professional bureaucracy, the standing army, the centralized revenue system, the state-recognized church. The apparatus was, in part, the institutional technology for preventing another continental calibration crisis. The state censored publications, not always effectively, but consistently enough to slow the propagation of incendiary claims. The state licensed printers, imposing some version of the village correction mechanism onto the industrial medium by making printers legally responsible for what they produced. The state certified the authority of its clergy, the credentials of its professionals, the authenticity of its documents, each certification an attempt to restore the calibration that the print economy had disrupted.

It is important to be careful about this part of the story. The early modern state was not a solution to the calibration problem in any general sense. It was a correction calibrated to the specific failure that the religious wars had produced, applied within the jurisdictions the wars had carved out. It solved that failure. It did not solve the deeper architectural problem of how to maintain calibration in an amplifier economy. It solved it for one instance. When the next amplifier arrived, the electrical telegraph, the mass-circulation newspaper, the radio, each one produced its own calibration crisis, and each crisis was absorbed by institutional innovations that sat on top of the new medium and performed a correction function analogous to the one the state performed on the printing press.

This is the pattern I want the reader to have in view as we turn back to the agent stack. The institution that gets built after the collapse is built on top of the amplifier, not inside it. It is not an improvement to the amplifier. It is a layer above the amplifier that performs the correction the amplifier's internal architecture cannot perform. The Reformation did not produce a better printing press. It produced a set of political institutions that could operate a printing press economy without the printing press economy collapsing into war every generation. The modern state did not fix printing. It built a structure around printing that kept printing's failure modes contained.

The Return to the Agent Economy

The compute and inference layer of the agent stack is this book's printing press.

I mean this in a specific architectural sense, not as a loose analogy. The printing press took a cognitive practice, the production of theological claims, that had been locally bounded by correction mechanisms calibrated to pre-industrial throughput, and amplified the practice to a throughput that stripped out those correction mechanisms. The compute and inference layer of the agent stack takes a cognitive practice, the production of reasoning and recommendation, that has been locally bounded by correction mechanisms calibrated to human conversational throughput, and amplifies the practice to a throughput that strips out those correction mechanisms.

The specific mechanism is the same. The pre-print correction loop depended on the priest and the buyer being in continuing relationship, on the buyer's other relationships informing their estimate of the priest's reliability, on the community aggregating these individual estimates into a reputation that disciplined the priest's behavior over time. The pre-agent correction loop, where we had one, depended on the human decision-maker being in continuing relationship with the population their decisions affected, on the decision-maker's other activities informing their colleagues' estimate of their reliability, on the professional community aggregating these estimates into a reputation that disciplined behavior over time. Neither loop scales to the throughput of the amplifier. Neither loop can be restored inside the amplifier by making the amplifier faster or cheaper or more widely available. The loop, if it is to be restored, has to be restored outside the amplifier, in an institutional layer that sits on top of the amplifier and that operates at a scale and in a structure that can discipline the amplifier's outputs.

The calibration crisis of the indulgence economy did not arrive immediately when the printing press was invented. It took decades. The printers had to build the capacity to produce certificates in large volumes. The Church had to develop the practice of commissioning large print runs. The pardoners had to develop the skills and the networks to distribute the printed certificates across geographies the handwritten system had not reached. The rhetorical conventions that would eventually produce the most aggressive claims had to develop. The competing theological positions that would take advantage of the crisis had to be articulated, distributed, and read. The whole apparatus took nearly seventy years to come into the configuration that produced Luther's intervention, and then another century and more to resolve.

The agent economy is not seventy years into its amplifier phase. It is, at the time of this writing, perhaps three years in, if we date from the broad availability of capable foundation models, and perhaps one year in, if we date from the broad deployment of agentic systems in enterprise contexts. The calibration crisis has not arrived. The capacity for it has been assembled. The distribution channels exist. The rhetorical conventions that will produce the most aggressive claims have begun to develop, visible in the marketing language of the vendor ecosystem and in the deployment pitches the consulting firms are making to their enterprise clients. The competing positions that will eventually contest the current architecture are being articulated in the academic and policy literature, but they have not yet reached the stage at which they could catalyze a continental response.

The window the rest of the book will attend to is the window between now and the arrival of the crisis. In the printing-press case, the window was long enough that thoughtful observers at the time could see the cognitive architecture of the new medium and could, in principle, have recommended the institutional structures that would eventually be built. They did not. The institutional structures were built after the collapse, calibrated to the shape of the specific damage the collapse produced. The question this book is asking is whether the current window can be used differently, whether the institutional layer that the agent economy will eventually require can be built before the collapse that would otherwise force it into existence.

The next chapter takes up the same structural pattern in a different medium, with a different pillar of the framework at its center, and a different historical population that paid the cost of the failure. The medium is the reporting chain. The pillar is Outcome Achievement. The population is the Chinese countryside in the years 1959 through 1961.

Chapter 4: The Reporting Chain

What Gets Counted Upward

A harvest is a measurable thing. It is grain, in specific quantities, stored in specific places, attributable to specific fields. The measurability is one of the oldest features of agricultural civilization. States have been measuring harvests for as long as states have existed, because harvests are the base on which states extract the surplus that makes statehood possible. The technology of harvest measurement, granaries, scales, census takers, crop inspectors, evolved across millennia in response to the persistent state interest in knowing what the countryside had produced.

The reporting chain that moves a harvest figure from the field to the capital is, likewise, an old institution. The field yields a number. The number is reported to a local official. The local official aggregates the numbers from their jurisdiction and reports to a higher official. The higher official aggregates further and reports upward. The chain terminates in some office that needs the national figure for some purpose, planning, taxation, grain procurement, famine relief. At each step, an aggregation happens, and at each step the number is a product not only of the harvest below but of the incentives that operate on the reporter.

The incentives are the point of this chapter. The harvest is a cognitive object: a representation of agricultural reality that the state carries in its ledgers and that the state uses to decide what to do. The reality of the harvest is what happens in the fields. The relationship between the cognitive object and the reality is the thing the framework calls Outcome Achievement. When the two correspond, the state can convert its figures into actions that produce outcomes in the countryside. When the two diverge, the state is making decisions about a representation of the countryside rather than about the countryside itself, and the decisions produce outcomes only to the extent that the representation happens to track the reality. When the divergence grows large, the decisions produce outcomes that bear no intended relationship to the countryside at all, and the countryside's actual state becomes visible only through events severe enough to be noticed outside the reporting chain.

The case this chapter examines is the most severe instance of this failure mode in the historical record, not because the underlying cognitive architecture was unusual, but because the amplifier in question was configured to maximize the decoupling between the representation and the reality. The amplifier was the totalitarian reporting chain of the People's Republic of China during the Great Leap Forward, and the decoupling it produced killed, by the most defensible demographic estimates, tens of millions of people across three years. The figures are estimates because even now, with seven decades of archival work, the exact magnitude of the famine is disputed. The central finding of that work is not in dispute: the grain figures that traveled upward through the reporting chain during 1958 and 1959 had, by the end of the sequence, become a pure fiction, constructed by the successive certification of lower-level fictions, and the state made decisions on that fiction that produced the largest peacetime mortality event of the twentieth century.

The Architecture of the Chain

The Great Leap Forward, launched in 1958, was an ambitious attempt to accelerate China's industrialization by mobilizing the rural population into a set of economic campaigns, agricultural collectivization at a scale beyond what had preceded it, mass steel production in backyard furnaces, vast construction projects, and above all the radical reorganization of the countryside into people's communes. The campaign was driven from the top by a leadership that had committed itself publicly to rapid and dramatic gains in grain production. The public commitment was consequential, because it made the reality against which lower-level performance would be judged the leadership's announced targets, not the empirical harvest.

The reporting structure that translated leadership targets into local quotas had, formally, the shape of any bureaucratic hierarchy. The village cadre reported to the production brigade. The brigade reported to the commune. The commune reported to the county. The county reported to the provincial authority. The provincial authority reported to Beijing. At each level, the report was a number: the grain output of the unit, expressed in standard agricultural units, aggregated from the numbers of the units below. The structure was, on paper, a straightforward information-gathering apparatus of the kind that modern bureaucracies have used for centuries.

The feature of the structure that mattered for the failure was not the hierarchy itself. It was the combination of the hierarchy with three other features that the Great Leap Forward configuration supplied.

First, the targets were set from the top and pushed downward, rather than derived from the bottom and aggregated upward. When Beijing announced a goal of doubling grain production in a year, the announcement was not a forecast. It was a commitment. The provincial authorities translated the commitment into provincial quotas. The counties translated the provincial quotas into county quotas. The communes translated those into commune quotas. By the time the quota reached the village cadre, it was a specific number that the village was expected to produce. The number had no origin in the village's actual agricultural capacity. It had an origin in the summation constraint that each level faced, which was that the reports going upward had to reconcile with the targets that had come downward.

Second, the political environment made the certification of target compliance professionally and sometimes physically imperative. The Great Leap Forward was, in its self-presentation, a revolutionary campaign. The political apparatus understood itself to be engaged in transforming the countryside, and it understood resistance to the transformation as counterrevolutionary. Cadres who reported shortfalls against the targets were vulnerable to accusations of sabotaging the campaign. Cadres who certified compliance were, for the duration of the campaign, politically safe. The incentive structure therefore rewarded upward certification regardless of ground truth and punished truthful reporting of shortfalls. The rationality of the individual cadre, under the incentive structure, was to certify.

Third, the informal correction mechanisms that might have surfaced the divergence between reports and reality had been systematically suppressed. Independent journalism, private correspondence across regions, visits by outside observers, the complaint systems that even traditional Chinese bureaucracy had maintained for centuries through the office of the censor, all of these had been weakened or dismantled by the preceding decade of revolutionary consolidation. The village peasant who knew that the reported harvest bore no relation to the actual harvest had no channel through which that knowledge could travel upward outside the reporting chain itself. The chain was, effectively, the only information about the countryside that the state received. And the chain was the specific mechanism that had been configured to report what the state wanted to hear.

Under these three conditions, the reporting chain became an amplifier in the precise sense this book uses the term. The individual cadre's decision to certify upward was a small act of miscalibration, analogous to the individual medieval priest's small act of theological slippage in the indulgence economy. The cadre's certification then became an input to the next level's aggregation, which was itself a certification, which became an input to the next level's aggregation. Each level's certification carried forward the miscalibration and, because the higher levels were themselves under pressure to produce numbers that reconciled with the targets above them, added additional miscalibration at each step. The amplification factor was large. By the time the provincial figures reached Beijing, they bore no systematic relationship to the grain that existed in the fields. They had become, as a matter of information content, a fiction.

The Collapse in the Fields

The consequences of the fiction were, for a period in 1958 and early 1959, invisible at the top. The Beijing leadership received reports of record harvests. The reports were consistent with their targets. The consistency reinforced the leadership's confidence in the campaign. On the basis of the reported production, Beijing set the state grain procurement quotas, the amount of grain the state would extract from the countryside for urban consumption, industrial production, and export in exchange for foreign currency. The procurement quotas were calculated as a share of the reported harvest. Because the reported harvest was much larger than the actual harvest, the procurement quotas took a much larger share of the actual harvest than the state had intended. In some regions, the quotas exceeded what the countryside had actually produced. The state extracted grain the countryside did not have.

The mechanism is worth following through the specific human consequences, because the specificity of those consequences is what makes the case an architectural study rather than a statistical curiosity. A village that had, in reality, produced enough grain to feed its population through the winter had, on paper, produced much more. The state took the procurement share based on the paper figure. The physical grain that left the village was the physical share of the paper harvest, not of the actual harvest. The grain that remained was what the paper arithmetic said should remain, minus what the paper said should have been extracted. When the paper arithmetic was far enough from the reality, the remainder was too small to feed the population. The population began to starve.

The onset of famine in the countryside was not an event the reporting chain could see. The chain's inputs were the harvest figures, not the nutritional status of the population. The political incentive to certify upward, which had produced the inflated harvest figures, now produced an incentive to suppress reports of starvation, because the appearance of starvation in a region that had just reported record production would immediately expose the preceding reports as fiction and the cadres who had produced them as liars. The cadres, under the same rationality that had produced the initial inflation, concealed the famine's early stages. They did this not because they were uniquely evil but because the system had configured their incentives such that concealment was the dominant strategy at the level of the individual cadre.

The concealment could only work for so long. When the death toll in a village reached certain thresholds, the fact became impossible to hide from neighboring villages, from the local medical personnel who had to deal with the bodies, from the rural officials who, whatever their political commitments, could no longer pretend that nothing was happening. Information about the famine began to leak upward through channels that were not the official reporting chain. It reached provincial officials who had the courage or the local independence to report it to Beijing. It reached party members who visited the countryside on other business and returned with accounts that could not be squared with the official figures. It reached the leadership, eventually, not through the system that was supposed to tell them what was happening but through the accidental channels that survived the suppression of the formal ones.

The central leadership's response to the emerging evidence is one of the most studied and most disputed episodes of twentieth-century political history. At the Lushan Conference in July 1959, Marshal Peng Dehuai submitted a letter to Mao Zedong raising concerns about the excesses of the Leap and the signs of developing difficulty in the countryside. The letter was a model of tactful criticism within the conventions of party communication, phrased in terms that preserved the campaign's ideological framing while questioning its specific execution. Mao's response was to circulate the letter to the full conference and to orchestrate a political campaign against Peng that ended with Peng's removal from his positions. The signal that Lushan sent through the party was that criticism of the Leap, even the most carefully phrased criticism, was politically impermissible. The reporting chain's inflation of harvest figures, which had been under pressure throughout 1958 and early 1959, became newly entrenched. The famine, which by this point was well underway in several provinces, intensified.

The period from the second half of 1959 through 1961 is when the bulk of the mortality occurred. The precise figures remain a matter of scholarly dispute, partly because the record-keeping during the period was itself corrupted, partly because the Chinese state has not fully opened its archives for independent analysis. The range of serious estimates runs from roughly fifteen million excess deaths at the low end to forty-five million at the high end, with a central range around thirty to thirty-five million. These are demographic estimates derived from population registration data, census comparisons, and specific regional studies. They are not political claims. They are the best reconstruction scholarship has been able to produce of what actually happened to the rural population during the famine years. The magnitude makes the episode the largest famine in recorded history by a substantial margin.

What matters for the argument of this book is that the scale of the mortality was not produced by any feature of Chinese agriculture, Chinese geography, or Chinese culture that was unique to the period. It was produced by the specific configuration of the reporting chain, operating under the specific incentives the political environment supplied, running for the specific period during which the central leadership was unwilling to revise its cognitive model of the countryside. The same mechanism, in a different polity with different political vocabulary, would have produced the same failure mode. The mechanism is not a property of totalitarianism in any essential sense. Totalitarian political structures configure the mechanism in its most extreme form, but the mechanism itself, the certification of upward-flowing reports under incentives that reward compliance over accuracy, is present in many bureaucratic systems and can produce analogous failures at smaller scales and with different specific casualties.

The Correction That Was Stripped Out

The correction mechanism that the Great Leap Forward configuration stripped out was the lateral feedback loop that, in most bureaucratic systems, allows different parts of the apparatus to cross-check each other's reports. A well-functioning bureaucratic chain is not a simple hierarchy. It is a hierarchy embedded in a lateral network of institutions that can observe the same reality through different channels and can compare their observations. A provincial governor reports harvest figures to the capital. The national grain board, which deals with the physical movement of grain, has its own information about how much grain is being transported, milled, and distributed. The statistical bureau, which conducts surveys on its own authority, has its own information about rural conditions. The press, even a controlled press, reports on economic conditions in ways that can be cross-referenced with official figures. The academic establishment, which studies agricultural productivity, has its own models that generate expected figures against which reported figures can be tested.

The lateral cross-checking is the bureaucratic analog of the village-scale correction mechanism in the indulgence case. It does not require the lateral institutions to be independent in any strong sense. It requires only that they have different immediate reporting lines, different specific incentive structures, and different organizational cultures, such that they are unlikely to fail in correlated ways on the same question. When one reporting line produces a figure that the others cannot square with their own observations, the inconsistency becomes visible to someone in a position to investigate. The investigation may or may not succeed, but the lateral architecture creates at least the possibility that the divergence between the reported figure and the reality can be surfaced before the reporting figure becomes the basis for catastrophic action.

The Great Leap Forward configuration had systematically weakened each of these lateral correction points. The grain board's reports were produced under the same political pressure as the agricultural reports. The statistical bureau had been effectively absorbed into the planning apparatus. The press had been disciplined into uniform coverage of the campaign's successes. The academic establishment had learned, through the anti-rightist campaign of 1957, that independent economic analysis carried serious personal risk. The result was that the reports traveling up the agricultural reporting chain were not subject to the lateral cross-checking that would have been, in a differently configured bureaucracy, the mechanism most likely to have surfaced the divergence.

This is what I mean by stripping out the correction system. The correction mechanism had not been abolished by any explicit decision. It had been made non-functional by the configuration of incentives across the institutions whose lateral operation constituted the correction. Any single institution whose reports could have contradicted the agricultural chain could have been silenced individually without any explicit policy to that effect, because the individuals who ran the institution had learned from the prior decade of campaigns that producing reports inconsistent with central party lines was hazardous. The stripping-out was emergent, in the technical sense that it was produced by the aggregation of individual responses to a political environment, not by any central policy of stripping.

This emergent character of the failure is, I think, the feature of the case most relevant to the argument about the agent stack. The Great Leap Forward is often remembered as a failure of central planning, and there is a sense in which that description is true. The failure's deeper architecture was not, however, a failure of the central planners' judgment. It was a failure of the information system that produced the inputs on which the central planners were operating. The planners were working with a picture of the countryside that the reporting chain had constructed for them, and the construction had been shaped by incentives that no single actor had chosen but that all the actors were responding to. The system as a whole was producing decisions that, from the inside, appeared to be based on the available evidence. The evidence itself had been corrupted by the same system that was using it. There was no outside view available to the people running the system, because the system had, over the preceding decade, configured itself to exclude the channels through which outside views might have entered.

The Institution That Got Built

The institutional response to the Great Leap Forward failure is one of the most consequential and least understood episodes of twentieth-century economic history. It did not arrive as a single reform. It arrived in stages across twenty years, culminating in the economic program that Deng Xiaoping initiated in the late 1970s and that transformed China's relationship to its countryside and, through that transformation, the structure of the global economy.

The immediate post-famine response, in the early 1960s under the leadership of Liu Shaoqi and Zhou Enlai, was a set of specific corrections to the agricultural reporting system. Quotas were lowered to realistic levels. The commune structure was relaxed to allow more household-level agricultural activity. The central statistical apparatus was reconstituted to some degree, although the political constraints on its operation persisted. These corrections were partial and were themselves reversed during the Cultural Revolution, which is one of the reasons the full institutional response took another decade to emerge.

The deeper response, which arrived with the economic reforms of the late 1970s and 1980s, was the reintroduction of market signals as a cross-check on the planning apparatus. The household responsibility system, which restored incentive-compatible agricultural production by allowing households to sell surplus on the market, was not only an economic reform. It was an information reform. A household that could sell its surplus on a market had an incentive to produce accurate information about what its surplus actually was. The market price of grain became an independent signal of actual grain availability, independent of what the reporting chain was certifying. Over time, the combination of market signals and a gradually reconstituted statistical apparatus produced a picture of the Chinese countryside that was considerably closer to the reality than the picture the planning apparatus had been operating on during the Leap.

The institutional innovation, in short, was the layered introduction of a correction mechanism that could operate on the reporting chain's outputs and check them against an independent reality. The market was, in this analysis, not primarily an economic device. It was an information device. It created the lateral feedback loop that the totalitarian configuration had stripped out.

The reform was calibrated, like the modern state's response to the printing press, to the specific failure that had preceded it. It did not attempt a general solution to the problem of information integrity in large bureaucracies. It addressed the specific mechanism by which the Leap had produced its catastrophe, which was the monopoly of the political reporting chain over information about the countryside. It addressed that mechanism by introducing an alternative channel, the market, whose structure produced information that could not be easily coordinated with the political channel's certifications.

The Return to the Agent Economy

The evaluation and observability layer of the agent stack is this book's reporting chain.

The mechanism is the same. An agent produces an output. The output is assessed by an evaluation module that scores it against quality metrics. The score is reported to the layer above. The layer above aggregates scores across many outputs and reports to the layer above it. The layer above aggregates across many workflows and reports to the enterprise dashboard. The enterprise dashboard reports to the executive who is deciding whether to expand the agent's deployment, adjust the quality bar, or intervene in the system's behavior. At each step, the report is an aggregation. At each step, the reporter has incentives to certify upward, because upward certification is what keeps the investment thesis intact, what sustains the vendor contract, what allows the deployment to continue into the next quarter. At each step, the correspondence between the reported score and the outcomes the enterprise actually cares about depends on whether the quality metrics, as operationalized, are tracking the outcomes. When they do, the chain works. When they do not, the chain is certifying outputs that bear an increasingly loose relationship to the outcomes.

The structural analogy is not rhetorical. It is architectural in the same sense. Both systems are reporting chains in which each level's output is the next level's input, each level has incentives to certify compliance with the targets above it, and the lateral correction mechanisms that might cross-check the certifications have been systematically absented from the configuration. In the Great Leap Forward case, the absence was produced by political pressure that suppressed lateral institutions. In the agent stack case, the absence is produced by the architecture of the layered vendor ecosystem, which has not yet developed the lateral institutions that would perform the cross-checking, because those institutions have not yet been either commercially viable or strategically required.

The specific failure mode the agent stack's reporting chain is most vulnerable to is the gradual decoupling of quality metrics from enterprise outcomes. The metrics were defined at the time of deployment, using the best available understanding of what the system was for. The metrics reflected that understanding. Over the following months and years, the environment in which the enterprise operates shifts. The outcomes the enterprise cares about shift with it. The metrics, however, are embedded in the evaluation infrastructure, in the contracts with the vendor, in the dashboards the executive team reads. The metrics do not shift at the same rate, because shifting them requires updating the infrastructure, renegotiating the contracts, rebuilding the dashboards. The metrics lag. For the period of the lag, the system's reporting chain is certifying that the system is doing well, when the system is doing well at a task that has ceased to be the task the enterprise needed it to do.

The specific enterprise domains where this failure mode is about to arrive will be the subject of Chapter 8. For now, it is enough to mark the architectural point. The evaluation and observability layer of the agent stack performs the function of the Great Leap Forward's reporting chain: it certifies outputs to the layer above without measuring outcomes. The specific failure of outcome achievement that this certification produces is the same failure the Chinese countryside lived through in 1959 and 1960, scaled to the economic consequences of a different kind of decision and a different kind of institution, but with the same structure. The institutional response that will eventually be built will be analogous to the response that was eventually built in China: the introduction of an independent channel that can cross-check the certifications of the reporting chain and surface the divergence between reported outputs and actual outcomes before the divergence becomes large enough to produce catastrophic decisions on the basis of fictional data.

The analogy should not be stretched beyond what the argument supports. The consequences of the Great Leap Forward famine were catastrophic in ways that the consequences of an agent economy miscalibration are unlikely to be, for reasons that have to do with the specific domains in which agents are deployed and the specific populations on which those deployments act. What the analogy does support is the claim that the architectural pattern is the same, that the pattern produces failures of a particular kind, and that the institutional response that will eventually arrive has a particular shape. The shape is the introduction of lateral correction. The introduction is difficult, because it requires building institutions that have incentives and capabilities distinct from the reporting chain's incentives, and because the vendors who operate the reporting chain are not the natural builders of those institutions. The introduction is, however, the direction in which the corrective pressure will eventually run, if the argument of this book is right about the shape of the failure. The question is only whether the introduction will happen before the specific scandals that will force it, or after.

The next chapter turns to a case in which the amplifier operated not through a political reporting chain but through a financial infrastructure, and in which the pillar of the framework most centrally implicated was not Outcome Achievement but Informed Navigation. The case is recent enough that most readers will remember it, which creates both advantages and hazards for the analysis. The advantages are that the evidence is extensive and the consequences are well-documented. The hazard is that familiarity can make it seem like the pattern I am describing is specific to financial markets. It is not. The pattern is architectural, and it recurs across media. The financial case is one instantiation. The agricultural case is another. The printing case is a third. The agent stack is the fourth.

Chapter 5: Bad Loans and Securitization

The Loan Officer Who Knew the Borrower

Before the era of securitization, a mortgage was a transaction between two parties who knew something about each other. A borrower walked into a bank. A loan officer reviewed the borrower's income, employment, credit history, and the property the loan was meant to finance. The loan officer made a judgment. The bank made the loan. The bank kept the loan on its own balance sheet. If the borrower defaulted, the bank took the loss.

This arrangement created a specific kind of cognitive environment for the loan officer. The officer had to care about the quality of the loan, because the loan was going to live with the bank, and the officer's own performance, compensation, and career depended on the aggregated performance of the loans the officer had approved. The officer's judgments were, in the framework's vocabulary, subject to Informed Navigation in the operational sense: the officer had specific information about specific borrowers, specific properties, specific local labor markets, and had to use that information to distinguish loans that would perform from loans that would not. The officer had access to the specific kinds of knowledge that generic models could not capture, the reputation of a local employer, the stability of a particular neighborhood, the character of a particular borrower, and was expected to weight that knowledge in the decision.

The system was not always accurate. Loan officers brought their own biases to the work. Local practices varied across institutions and geographies. The aggregate loss rates of the pre-securitization mortgage system were not zero, and the system's errors fell along the predictable dimensions that bias in loan evaluation has always produced. What the system did have, however, was the structural property that the cognitive apparatus making the navigation decision was in direct contact with the information on which good navigation depended. The officer knew what the officer did not know, and could weigh that not-knowing into the decision, because the officer was the one who would bear the consequences of the decision.

This arrangement is worth holding in mind, because the shift that the chapter traces was not a shift from a good system to a bad one. It was a shift from a system in which the navigation function was performed at the point of information to a system in which the navigation function was distributed across a series of actors, each of whom had progressively less of the underlying information and progressively more of the appearance of sophisticated analysis. The result was that the cognitive weight of the decisions moved further from the information on which the decisions depended, and at the point of greatest distance, the decisions were being made on the basis of abstractions whose correspondence to the underlying reality had become a premise rather than a fact.

The Amplifier

Securitization is a set of financial techniques that convert pools of individual loans into tradeable securities. The basic idea is old. The specific implementation that produced the 2008 crisis was comparatively new, and the newness matters, because the features of the new implementation were precisely the features that stripped out the correction mechanisms of the pre-securitization system.

The basic structure of a mortgage-backed security is straightforward. A set of individual mortgages is pooled into a trust. The cash flows from the mortgages, the principal and interest payments the borrowers make each month, flow into the trust. The trust issues securities whose payments are derived from those cash flows. An investor who buys one of these securities is, in effect, buying a share of the payments from the underlying pool of mortgages.

The simplest version of this structure, the pass-through security, simply divides the cash flows pro rata among the investors. An investor who owns one percent of the trust receives one percent of the monthly cash flows. The risk of the pool, the probability that some fraction of the borrowers will default, is borne proportionally by the investors. The structure creates a market for mortgage credit, which, in principle, allows capital to flow to housing finance more efficiently than it would if each lender had to keep each loan on its own books.

The complication, and the place where the Informed Navigation failure entered the system, was the innovation of structured securitization: the division of the cash flows into tranches with different claims on the pool's payments. A senior tranche has first claim on the cash flows and is paid in full before anything flows to the junior tranches. A mezzanine tranche is paid after the senior tranche but before the equity tranche. The equity tranche is paid last and absorbs the first losses from defaults in the pool.

The structural innovation did something remarkable to the analytical problem of valuing the securities. The pass-through security's value depended on the default behavior of the pool as a whole, which in turn depended on the specific features of the underlying loans. Evaluating the security required evaluating the pool. The tranched security's value depended on different aspects of the pool's behavior for different tranches: the senior tranche cared about the probability of catastrophic pool losses, the mezzanine tranche cared about the distribution of moderate losses, the equity tranche cared about expected losses. Different statistical features of the pool determined the value of different tranches. The evaluation of each tranche could be performed using a model that captured the relevant features. And once the model was specified, the evaluation of the tranche did not require any direct engagement with the individual loans.

This is the point at which the Informed Navigation function was handed off. In the pass-through system, the investor was still exposed to whatever the individual loans were actually doing, and a sophisticated investor had reason to ask questions about what those loans actually were. In the tranched system, the senior investor was told, by the structure of the security and by the rating agency's evaluation of the structure, that the probability of the catastrophic pool losses that would affect their tranche was negligible. The negligibility depended on the model used to evaluate the pool. The investor, in purchasing the senior tranche, was trusting the model. The investor's own Informed Navigation was over the space of tranches and rating categories, not over the space of loans. The distance between the investor's cognition and the underlying information had been extended by one step.

The Correction That Was Stripped Out

The correction mechanisms of the pre-securitization mortgage system, like the correction mechanisms of the pre-print indulgence economy, depended on proximity between the decision-maker and the information. The loan officer knew the borrower. The bank that kept the loan knew the loan officer. The regulator that supervised the bank knew the bank's local market. Each link in the chain was short enough and local enough that the information necessary for good navigation could flow along the link. When the bank made a bad loan, the loss landed on the bank's own balance sheet, which produced feedback to the loan officer, which produced feedback to the officer's subsequent decisions.

The securitization system broke each of these links in a specific way.

The first broken link was between the loan officer and the consequences of the loan. In the traditional system, a bad loan was the officer's problem. In the securitization system, the loan was sold to the securitizer shortly after origination. The originating institution no longer bore the loss. The officer's performance was measured by the volume and documentation of loans originated, not by the subsequent performance of those loans. The incentive to use the officer's specific, local, informal knowledge of borrowers to distinguish good loans from bad ones weakened, because the officer was no longer bearing the cost of failing to do so. The incentive to originate in volume strengthened, because volume was what the compensation system rewarded. The officer still had the local information, but the system no longer called on the officer to use it.

The second broken link was between the originator and the pool. The institution that originated the loans sold them to the securitizer. The securitizer aggregated loans from many originators into a pool, structured the pool into tranches, and sold the tranches to investors. The securitizer did not have the originator's local information. The securitizer had the originator's loan files, which contained the documentary record of the loan. The documentary record was, however, what the securitizer could work with, and the documentary record did not capture the informal judgments the loan officer had previously been expected to make. The securitizer's evaluation of the pool was a statistical exercise over the documentary features of the loans, not an informed judgment about the quality of the loans.

The third broken link was between the pool and the rating agency. The rating agency's job was to evaluate the probability that each tranche of the security would experience losses at different levels. The evaluation was a model-based exercise: a statistical model of the pool's likely default behavior, parameterized by the pool's documentary features, calibrated against historical loss rates for similar pools. The rating agency did not have access to the pool's individual loans in a way that would have supported direct engagement. It had aggregate pool characteristics and a model that translated those characteristics into expected losses. The model was the agency's navigation. The investor who bought the rated tranche was, in turn, trusting the agency's model.

The fourth broken link was between the rating agency and the investor. The investor who bought the senior tranche of a mortgage-backed security was typically a large institutional investor, a pension fund, a sovereign wealth fund, a bank treasury, whose investment mandate required the investment to carry a specific credit rating. The mandate-setting process treated the rating as a reliable summary of the security's risk. The investor did not perform an independent evaluation of the underlying pool. The investor's navigation was over the space of rated securities, not over the space of mortgages. When the investor's mandate said "AAA rated securities," the investor's due diligence consisted of confirming the rating, not of revisiting the model that had produced it.

Each of these four breaks removed an opportunity for local information to enter the valuation chain. The officer's knowledge of the borrower did not reach the securitizer. The securitizer's knowledge of the pool did not reach the rating agency in the form that would have supported model revision. The rating agency's model did not surface its own assumptions to the investor. The investor's judgment did not extend below the rating. The Informed Navigation function, which the pre-securitization system had performed at the loan-origination step, had been distributed across a chain of actors in which no actor had the combination of information and incentive to perform it.

This was the specific form the Informed Navigation failure took. It was not that any single participant in the system was incompetent. Each participant was working within the frame their role defined. The loan officer originated loans the bank's policies authorized. The bank sold loans to the securitizer at prices the market supported. The securitizer structured pools according to the rating agency's guidelines. The rating agency applied its published methodology. The investor bought securities consistent with its mandate. No one was doing anything that any of the others would have criticized. The failure lived in the architecture of the chain, not in any single participant's behavior.

The Collapse

The specific collapse that followed is familiar enough in outline that the chapter can compress it. The premise on which the system had been operating, that the default behavior of the pools could be modeled using historical data from a period of rising housing prices, was false. The period of rising housing prices had been the data on which the models were calibrated. The assumption that the period would continue was the premise the models rested on. When housing prices stopped rising and then began to fall, the default behavior of the pools diverged sharply from what the models had predicted. The divergence was not gradual. The default rates, which had been low and stable during the rising-price period, climbed steeply when the prices reversed, because many of the loans that had been originated during the late years of the rise had been made on the assumption that continued price appreciation would cover any weakness in the borrower's underlying capacity to pay. When the appreciation reversed, those loans began to default together, in correlated waves across geographies that the models had treated as independent.

The tranched structure of the securities, which had been designed to protect senior investors from the behavior of a pool's underlying loans, did not work as designed when the pool's losses exceeded what the model had anticipated. The equity and mezzanine tranches were wiped out quickly. The senior tranches, which had been rated as near-risk-free and sold to investors whose mandates required safety, began to experience losses that, by the structure's own terms, should not have been possible. The rating agencies downgraded the securities in waves, which forced institutional holders to sell into a market that was no longer willing to bid for the downgraded paper at any price close to its previous valuation. The liquidity crisis that followed, compounded by the opaque network of counterparty exposures among the financial institutions that had originated, securitized, held, and insured the securities, produced the systemic financial collapse whose formal date is September 2008 but whose duration extended across the following eighteen months and whose long-term consequences are still being processed by economies around the world.

The casualty structure of the 2008 crisis was different from the casualty structures of the indulgence crisis and the famine. No one starved. The armed conflicts that followed were limited. What happened instead was a sustained economic contraction that reduced the employment, savings, and future prospects of hundreds of millions of people who had no direct exposure to the securities that had collapsed. The unemployment rate in the United States rose from roughly five percent in 2007 to ten percent in 2009 and did not return to the pre-crisis level for several years. Comparable or more severe effects appeared across the European economies that had exposure to the same assets. The mortgages that had been the raw material of the crisis went into foreclosure at a rate that displaced millions of families from their homes. The fiscal responses required to stabilize the financial system produced sovereign debt loads that continue to constrain public policy across most of the developed world. The distribution of the costs was, as is typical of systemic financial crises, regressive: the populations that had been most marginally situated before the crisis bore the most durable consequences after it.

The Institution That Got Built

The institutional response to the 2008 crisis followed, with some specific modifications, the pattern this book has been tracing. Reforms were introduced that operated on top of the securitization infrastructure, performing correction functions the infrastructure could not perform for itself.

The Dodd-Frank Act, passed in 2010, was the centerpiece of the American response. Its provisions ranged widely, but the ones most relevant to the Informed Navigation failure were the provisions that required originators to retain a portion of the credit risk on the loans they securitized, the provisions that expanded disclosure requirements on the composition of securitized pools, and the provisions that restructured the role of the rating agencies to reduce the conflicts of interest that had shaped their pre-crisis incentives. Each of these provisions was an attempt to reinsert into the system an element of the cognitive function the chain had stripped out. The retention requirement restored a version of the originator's skin in the game. The disclosure requirement restored a version of the investor's access to the underlying information. The rating agency reforms attempted to restore a version of the evaluator's independence from the market being evaluated.

The Basel III framework, which was negotiated internationally and implemented across major jurisdictions through the decade following the crisis, addressed the crisis from a different angle. It required banks to hold more capital against their asset exposures, to hold specific reserves against the liquidity risks their balance sheets carried, and to conduct stress tests that would surface the specific vulnerabilities of their portfolios to scenarios the pre-crisis models had not adequately captured. The capital requirements and stress tests did not fix the Informed Navigation failure. They put a buffer around it. The buffer absorbed the cost of the next navigation failure, when it arrived, and slowed the propagation of that failure through the financial system.

The stress tests are the part of the response most relevant to the argument of this book. The stress test is, in its architecture, a specific kind of correction layer. It takes a bank's portfolio and subjects it to a set of adverse scenarios that were chosen by regulators on the basis of what the pre-crisis models had missed. The test asks: what happens to this portfolio if housing prices fall thirty percent, if unemployment reaches ten percent, if interest rates spike by three percentage points, if the commercial real estate market contracts sharply. The scenarios are not forecasts. They are plausible adverse conditions that the bank must demonstrate it could withstand. The stress test is, in effect, the introduction of an external cognitive check on the internal models of the institution. The bank's models tell the bank one story. The stress test tells the bank what the regulator thinks the model is missing.

The stress test has significant limitations. It only catches the failures regulators have imagined. It imposes a substantial compliance cost. It can produce gaming behavior, in which banks optimize for passing the test rather than for underlying safety. These limitations are real, and they are the subject of ongoing debate among the people who design and administer the tests. What the stress test does have, in its basic architecture, is the feature that defines a correction layer: it operates outside the system being corrected, and it applies a cognitive check the system cannot apply to itself.

The Return to the Agent Economy

The orchestration and multi-agent workflow layer of the agent stack is this book's securitization architecture.

The mechanism is the same in its architectural structure. A master query is decomposed into sub-queries. The sub-queries are delegated to specialized agents. Each specialized agent performs its sub-task against a specification that assumes the decomposition was sound and that the premise of the master query was correct. The results are composed into a recommendation. The investor in the recommendation is the user, who receives a synthesized output that does not surface the premises embedded in the decomposition. The synthesis has the same structural property that the senior tranche of a mortgage-backed security had: it appears to have integrated the information from the underlying components in a way that renders the underlying components irrelevant to the user's evaluation.

The specific failure mode that this architecture is most vulnerable to is the silent propagation of a wrong premise through a tree of sub-tasks. When the master query is framed correctly, the orchestration produces a reliable synthesis. When the master query is framed incorrectly, the orchestration produces a deeply internally consistent synthesis of wrong answers to wrong sub-questions. The user, who receives the synthesis, has no natural mechanism for interrogating the premise, because the premise is not surfaced in the output. The output looks like the result of a sophisticated multi-agent process. The user trusts the sophistication. The sophistication is real at the level of execution. The premise underneath the execution is what the user is being asked, implicitly, to trust.

The analogy to the securitization case is not that the agent stack will produce a financial collapse. It may or may not, depending on where the stack is deployed and what the specific consequences of the failures turn out to be. The analogy is that the architectural pattern of orchestration, in which a cognitive task is decomposed into subtasks that each inherit the decomposition's premises without being able to question them, is the same pattern that securitization instantiated in financial services. When the premises of the decomposition are sound, the system works. When the premises are not sound, the system produces failures that are hidden from the user by the very sophistication of the execution.

The correction layer that will eventually be built for the agent stack, if the historical pattern holds, will be analogous to the stress test in its architecture. It will be an external cognitive check that operates on the outputs of orchestrated agent workflows and applies adversarial scenarios designed to surface the premise failures the internal evaluation layer cannot surface. It will take a given workflow and ask: what happens to the output if the premise of the master query were different, if the data distribution underlying the specialized agents were different, if the coordination assumptions between agents were different. The scenarios will not be forecasts. They will be adversarial probes into the parts of the space the system has no independent mechanism for evaluating. The layer will not prevent premise failures. It will impose a cognitive buffer around them that absorbs the cost of failures when they arrive and slows their propagation through enterprise decisions.

What that correction layer looks like in practical terms, how it can be built, who builds it, who pays for it, on what timeline, is the subject of Chapter 11. For now, the point to retain is that the architectural requirement for such a layer is a direct consequence of the orchestration pattern, and the failure mode the layer would address is already visible in early deployments of multi-agent systems. The question is not whether the correction layer will be built. The historical pattern is unambiguous on that point: the correction layer always gets built after the collapse that establishes its necessity. The question is whether the current instance of the pattern will produce the correction layer before the collapse rather than after.

The next case involves an amplifier that was neither informational nor financial but administrative. It is the longest-running of the four, the most exhaustively studied, and the one most directly relevant to the memory-and-context layer of the agent stack.

Chapter 6: Without Market Feedback

The Problem That Central Planning Was Built To Solve

A modern economy produces, every day, an enormous quantity of small informational events. A bakery in Kiev runs out of flour at eleven in the morning. A shoe factory in Odessa discovers that its rubber supplier has shipped a lot of the wrong grade. A collective farm near Kharkov has an unexpectedly good wheat yield because the spring rains came at the right time. A mine in the Urals loses a week of production because a ventilation shaft needs repair. Each of these events is a fact about what the economy can produce and what it needs. In aggregate, the events constitute the economy's current state. In detail, they are what the economy's participants know about their own particular situations.

The cognitive problem of economic planning is to aggregate these events into a usable picture of what the economy is doing and what it needs. In a market economy, the aggregation is performed by prices. The flour shortage in Kiev produces a price pressure on flour, which signals to producers elsewhere that shipping flour to Kiev would be profitable, which brings flour in from wherever it can be most efficiently spared, which resolves the local shortage without anyone at the center needing to know why the shortage existed or how it was resolved. The price system is, from one angle, a distributed algorithm for aggregating local information into coordinating signals. It works, when it works, because the people who have the information also have the incentive to act on it, and the signals their actions produce reach everyone else at roughly the speed the original events occurred.

Central planning proposed a different solution. Rather than letting the aggregation happen through prices, the planners would collect information directly from the economic units, integrate it at the center, and issue production and distribution instructions to the units. The approach had certain theoretical appeals. It promised to eliminate the waste the planners saw in market economies: the unemployment that arose from demand shocks, the overproduction that arose from coordination failures, the inequalities that arose from the distributional properties of market outcomes. If the planners could produce a sufficiently accurate picture of the economy and could issue sufficiently precise instructions, the economy could be steered toward outcomes the market was unable to produce.

The theory required, as its condition of possibility, that the center could actually maintain a sufficiently accurate picture. This is the condition the Soviet experiment was a seventy-year test of, and the condition it empirically failed to meet. The failure, like the failures in the other historical cases this book examines, was not a failure of anyone's competence. The planners who ran Gosplan, the central planning apparatus, were in many cases talented economists with real analytical skills. The failure was architectural. The mechanism that the system used to produce its picture of the economy had, built into its structure, a specific reason to diverge from the economy it was supposed to picture, and the divergence accumulated over time in a direction that no amount of planning skill could reverse without changing the architecture itself.

The Amplifier

The Soviet planning apparatus was, in its essentials, a multi-layered bureaucracy that specified production targets for hundreds of thousands of enterprise units and specified deliveries among those units. The targets were expressed in physical quantities of outputs, tons of steel, square meters of cloth, kilowatt-hours of electricity, and in some cases in monetary terms that were themselves administratively set rather than market-determined. The deliveries were specified in material balances, matrices that listed what each enterprise was supposed to send to whom. The whole structure was updated on a five-year cycle, with detailed annual plans specifying the specific targets each enterprise had to meet each year.

The information that went into the plan came, in principle, from the enterprises themselves. Each enterprise reported, at the planning stage, what it needed in the way of inputs to meet its targets and what it expected to produce. The reports aggregated upward through the ministries that supervised each industry, through the regional planning offices, and finally into Gosplan, which integrated them into a coherent plan for the entire economy. The plan was then disaggregated back down through the hierarchy, with each unit receiving its specific targets and input allocations.

The architecture has a certain elegance on paper. The enterprises have the local information. The information flows upward. The center integrates it and issues instructions that are consistent with the aggregated reality. The problem is that the flow of information upward is not disinterested. The enterprise that reports its production capacity to its ministry knows that the reported capacity will become the basis for the target it is assigned. If the enterprise reports a higher capacity, it gets a higher target. If it reports a lower capacity, it gets a lower target. The enterprise's incentive is to report the capacity that will produce the target it can comfortably meet, which is lower than its actual capacity. The same calculation applies to every enterprise. The center therefore receives, across the entire economy, systematically understated capacity figures.

The planners were aware of this problem. They had been aware of it since the earliest years of the system. The response they developed was to add a margin to the targets that came back down. If the aggregated reports from the enterprises said the economy could produce some quantity of steel, the plan would assign targets that summed to a higher quantity, on the reasoning that the enterprises had understated their capacity and could produce more if pushed. The targets would be, in the planning vocabulary, taut: set above the reported capacity, so as to extract the additional effort that the enterprises were presumed to be withholding.

The taut targets produced their own response. The enterprises, assigned targets that exceeded what they had reported they could produce, had to find ways to meet the targets. One way was to actually increase production. Another way was to manipulate the reporting in ways that made the target appear to have been met without the underlying production having occurred. The methods for this manipulation were numerous. Enterprises could substitute lower-quality outputs for higher-quality ones, because the targets were specified in physical units and a ton of low-grade steel counted the same as a ton of high-grade steel. Enterprises could borrow outputs from future periods, reporting production that had not yet occurred against targets that were due now, on the implicit agreement that the borrowed production would be made up later. Enterprises could misclassify inputs and outputs to move quantities between accounts that were easier and harder to meet. Enterprises could, in the last resort, simply report production that had not occurred at all.

Each of these responses was, from the individual enterprise's perspective, a rational response to the incentive structure the taut targets imposed. Each response also introduced noise into the information the planners were receiving. The reports traveling upward contained some unknown mixture of actual production, substituted production, borrowed production, misclassified production, and fictional production. The planners had no reliable way to distinguish the components. They could observe that the aggregate reports generally matched the plan, which was the structural feature the reports had been shaped to produce. They could not observe whether the economic reality underneath the reports matched the plan.

This is the amplifier in this case. The planning apparatus amplified a particular cognitive failure, the inability of a central authority to distinguish compliant reports from accurate reports, across the entire economy, year after year, for seven decades. The amplification did not require anyone to be acting in bad faith. It required only that the enterprises respond rationally to the incentives the taut targets imposed, and that the planners' compensating adjustments maintain the taut targets regardless of whether the reports the enterprises submitted continued to reflect the underlying reality.

The Correction That Was Stripped Out

The correction mechanism that the planning system had stripped out was the mechanism by which, in a market economy, local information about real conditions reaches the center through the behavior of prices. The stripping-out was not a side effect of the planning system. It was the planning system's defining feature: the planners had explicitly replaced the price system with the reporting system, on the theory that the reporting system would carry the same information more efficiently.

The theory had a specific error. The price system carries information through the behavior of participants who have the information and who act on it in ways that adjust prices. The reporting system carries information through the behavior of participants who have the information but whose actions are structured by the incentives the reporting system itself creates. The two systems are not interchangeable carriers of the same information. They are different mechanisms that produce different informational properties. The price system's information is, to a first approximation, veridical about local conditions, because the participants who produce the price movements bear the consequences of acting on them. The reporting system's information is, to a first approximation, strategic about local conditions, because the participants who produce the reports have no incentive to make the reports veridical and strong incentives to make them serve the reporter's interests.

What the Soviet planners lost when they replaced the price system with the reporting system was the specific feature of market economies that keeps the central representation of the economy in contact with the local reality. In market economies, if the center holds a wrong picture of some part of the economy, the wrong picture produces decisions that fail in ways that are costly to the decision-makers, and the failures produce price movements that update the central picture. The updating is painful, slow, and imperfect, but it is real: the market signal eventually reaches the center, because the market's participants have the incentive and the capacity to act on the local information. In the Soviet system, there was no analog. If the center held a wrong picture of some part of the economy, the wrong picture produced decisions that failed in ways that were costly to the enterprises, but the enterprises had no mechanism for updating the center's picture. They could only comply with the wrong picture, while finding local workarounds that made the failure less painful at the local level.

The Resilience failure of the Soviet system was therefore a failure of the updating loop. The system could produce decisions, could issue plans, could move resources in accordance with the plans. What it could not do was revise its model of the economy against the accumulating evidence that the model was wrong. The evidence existed. It existed in shortages of specific goods in specific regions, in chronic over-production of certain outputs, in the black and gray markets that arose to redistribute goods the plan had misallocated, in the informal networks through which enterprises traded among themselves to compensate for the planning system's errors. These phenomena were visible to anyone who looked at the economy closely. They were not visible to the planning system, because the planning system was looking at its own reports, and the reports had been shaped to be consistent with the plan.

The Slow Collapse

The collapse of the Soviet system is conventionally dated to the political events of 1989 and 1991, but the economic collapse had a much longer arc. From roughly the mid-1970s onward, the Soviet economy was in a condition of accelerating stagnation that was visible in many indicators: slowing growth rates, widening gaps between published and estimated actual figures, chronic shortages of consumer goods, declining productivity in key industries, increasing dependence on hard-currency earnings from energy exports to cover the gap between what the domestic economy could produce and what the state needed to procure. Each of these indicators was, in isolation, a problem the planning apparatus could claim to be addressing. In aggregate, they were evidence that the planning apparatus had lost the capacity to update its model of the economy at the pace the economy was actually changing.

The planners were not unaware of the problem. Throughout the 1970s and 1980s, there were repeated attempts at economic reform within the planning framework: the Kosygin reforms of the 1960s, various experiments with enterprise autonomy in the 1970s, the perestroika reforms of the late 1980s. Each attempt tried, in its own way, to restore something like a feedback loop between the local economy and the central plan. Each attempt ran into the same structural problem. The feedback loop that the planners wanted was the feedback loop that a market economy produces through prices. The planners could not institute a price system without giving up the central authority that the planning system depended on. Each reform therefore offered partial feedback without full market signaling, and each reform ran up against the problem that partial feedback in a system that was otherwise unreformed produced distortions the unreformed system could not absorb.

The perestroika period is particularly instructive. Gorbachev's reforms granted enterprises additional autonomy in production decisions and allowed a limited form of private enterprise in certain sectors. The reforms were intended to introduce feedback into the planning system without dismantling it. In practice, the partial introduction of market mechanisms into a system that still operated under planning targets produced a set of incompatibilities that accelerated the system's breakdown. Enterprises that had been given autonomy used the autonomy to satisfy their own interests rather than the plan's interests. The plan's targets, which remained in place, became increasingly unmet. The reporting chain, which had been corrupted across decades, could not suddenly become a reliable source of information. The system as a whole began to malfunction in ways that the political apparatus had no mechanism for addressing, because the political apparatus was itself a product of the same system.

The formal collapse in 1991 was the political event that attended the economic collapse that had been underway for at least fifteen years. The Soviet state had lost the capacity to feed its urban populations reliably, to supply its industries with the inputs they needed, to maintain the living standards its population had come to expect, to project the military power its ideology required, to extract from its satellite states the tribute its economy depended on. Each of these capacities had been eroded by the same underlying failure: the planning apparatus could not update its model of the economy at the pace the economy required, because the reporting system that fed the model had long since ceased to carry accurate information, and no substitute information system had been constructed that could.

The human costs of the collapse are harder to summarize than the human costs of the other historical cases, because they were distributed across a longer period and a more complex set of consequences. The immediate costs in the early 1990s were severe: the post-Soviet transition produced declines in life expectancy, increases in mortality, and declines in living standards across most of the former Soviet territory. The longer-term costs included the institutional vacuum that made many of the post-Soviet states vulnerable to forms of governance that have proved, in several cases, more corrupt and less capable than what preceded them. The costs continue to be felt, in ways that make a final accounting premature.

The Institution That Got Built

The institutional response to the collapse of Soviet central planning was different from the responses in the other historical cases, because the collapse was not followed by the systematic reform of the existing system. It was followed by the attempt to construct a functioning market economy on the ruins of the planning apparatus. The attempt was more successful in some jurisdictions than in others, and the variation in outcomes is itself instructive.

The jurisdictions that made the most successful transitions, the Baltic states, Poland, the Czech Republic, were characterized by several features that allowed them to build market institutions relatively quickly. They had populations with living memories of pre-communist institutions. They had proximity to functioning market economies whose practices they could adopt directly. They had political leadership with coherent reform agendas. They had, perhaps most importantly, institutional capacity to enforce property rights and contract law at a level that allowed economic actors to plan on the basis of legal stability. The result, across the decades following 1991, was the construction of functioning market economies that today operate with Resilience properties that the planned system had not been able to produce.

The jurisdictions that made less successful transitions, Russia, Ukraine, several Central Asian republics, lacked various combinations of these features. Russia's transition in the 1990s produced a privatization process that concentrated much of the economy's productive assets in the hands of a small number of individuals with political connections. The resulting structure was neither the old planning system nor a functioning market economy but a hybrid in which the formal structures of the market economy operated within an informal structure that retained many of the features of the preceding system. The Resilience problem was partially addressed but not solved. The economy could update against local information to some degree, through the price system, but the price system operated within an institutional environment that limited the updating the prices could produce.

The institutional innovation that the successful transitions brought was, at its core, the reconstruction of the mechanism by which local information reaches the center through market prices. This is a narrower claim than the celebratory accounts of the transition period often made. The market is not a general solution to social problems. It is a specific mechanism for aggregating distributed information into coordinating signals, and its coordinating function depends on a set of institutional preconditions that are not automatically present. When the preconditions are present, the market does the updating that the planning system could not do, and the economy acquires the Resilience property that the planning system had lacked. When the preconditions are not present, the market produces information, but the information is distorted by the same institutional weaknesses that had distorted the planning system's information, and the updating is correspondingly compromised.

What matters for the argument of this book is that the institutional response to the planning failure took the form of reconstructing a feedback mechanism whose absence had been the core failure. The market was, in this analysis, not primarily an economic device. It was an information device. It created the updating loop that the planning system had stripped out.

The Return to the Agent Economy

The memory and context layer of the agent stack is this book's central plan.

The mechanism is the same in its architectural structure. An agent operates on a memory that encodes what the system has learned about its enterprise, its users, the problems it has been asked to solve, the patterns it has been rewarded for recognizing. The memory is not a direct correspondence with current reality. It is a compressed representation of prior reality, filtered through the interactions that produced it, shaped by the retrieval logic that governs which memories are surfaced in which contexts. When the reality shifts, the memory does not automatically shift with it. The memory retains the pattern it encoded, and the pattern shapes every subsequent retrieval.

The failure mode that this architecture is most vulnerable to is the progressive decoupling of the memory from the reality. The decoupling is produced by the same architectural feature that makes the memory useful: the memory is an aggregation of past interactions, weighted by some retrieval logic, and the aggregation is what allows the agent to generalize across situations rather than treating each query as if it had no context. The aggregation is also what makes the memory resist updating, because any specific piece of new information that contradicts the aggregated pattern is, by the memory's own logic, a single outlier against many confirming examples. The memory treats the outlier as noise. The noise accumulates. At some point, the outliers stop being outliers and become the new pattern, but the memory does not recognize this until the old pattern has produced a sufficient number of conspicuous failures that the aggregation tips.

The Resilience failure of the memory layer is therefore a failure of updating loop. The system can produce recommendations, can execute workflows, can synthesize outputs in accordance with the patterns its memory encodes. What it cannot do, except through mechanisms that are not yet built, is revise its model of the environment against the accumulating evidence that the model is wrong. The evidence exists. It exists in the specific queries that the system handles badly, in the specific outcomes that diverge from what the memory predicted, in the specific environmental shifts that human operators observe but that the memory has no channel for ingesting. These phenomena are visible to anyone who works closely with the agent. They are not visible to the agent, because the agent is looking at its own memory, and the memory has been shaped to be consistent with the patterns it has already encoded.

The correction layer that will eventually be built for this failure, if the historical pattern holds, will be analogous to the market in its architecture. It will be an external feedback mechanism that operates on the outputs of the agent's memory and produces signals that update the memory against current reality. What the Soviet case establishes is that such a mechanism is not a decoration on the system. It is a structural requirement. A system that operates on an internally consistent memory without an external updating loop will, across long enough time horizons, diverge from reality, because reality will not wait for the memory to revise itself. The divergence is not a bug in the system. It is a property of systems of this structural form. The Soviet planning apparatus diverged over seven decades. The agent stack's memory layer will diverge over whatever time horizon is determined by the rate of change in the environments the stack is deployed into and the inertia of the specific memory mechanisms the stack employs.

The four historical cases this part of the book has examined establish, I hope, the pattern clearly enough that the remainder of the argument can proceed on the assumption that the pattern is real. The pattern is that civilizations construct amplification media that strip out the informal correction systems that had kept a cognitive failure mode locally contained, that the stripped-out correction systems are visible only in retrospect, that the eventual institutional response takes the form of a new correction layer built on top of the amplifier rather than inside it, and that the lag between the amplification and the correction is measured in the casualties the intervening collapse produces. The indulgence case took a century. The Great Leap Forward took three years. The 2008 crisis took eighteen months. The Soviet collapse took seven decades.

The agent economy is the current instance. The question the rest of the book takes up is not whether the pattern will repeat. The pattern will repeat. It will repeat unless something specific is done to interrupt it. The question is what that something is, how it is built, and whether the civilization currently constructing the agent economy will build it before the collapse or after.

We turn next to the theoretical chapter that names the architectural principle underneath all four of the historical cases, and to the specific enterprise domains in which the pattern is most likely to arrive first.

Part III: The Agent Economy in Depth

The historical chapters establish that the failure mode is not speculative. It has run four times, across centuries and continents, through amplifiers as different as mechanical type, totalitarian hierarchy, financial securitization, and the Soviet planning apparatus. The pattern held across all four instantiations, which means the pattern is not a contingent feature of any specific technology or political system. It is a property of what happens when a familiar cognitive failure is distributed through a medium that outpaces the correction systems that used to contain it.

Part III returns to the present. The first chapter names the architectural principle underneath the pattern, the principle that makes the historical cases more than a collection of bad outcomes that happen to share a narrative shape. The second walks through the specific enterprise domains where the agent economy is already making or about to make decisions on uncalibrated premises, and traces what the collapse will look like in each. The third takes the risk of prediction, because a book that claims the pattern is legible in advance has an obligation to say what it expects to see in the next eighteen to twenty-four months. The predictions will be right or wrong on a timeline that makes the book testable. That is the point.

Chapter 7: The Scale Signature

Every correction system that humans have built has a throughput. By throughput I mean the rate at which the correction system can process the inputs it was designed to correct. A village priest who hears confession can correct perhaps several dozen parishioners a week. A bank underwriter who reviews loans can evaluate perhaps several hundred a year. A supervisor who reviews subordinates' reports can scrutinize perhaps several dozen per reporting cycle. A regulator who audits financial institutions can complete perhaps several dozen examinations per year per examiner. In each case, the correction system has a rate at which it can perform the correction, and the rate is a function of the cognitive, institutional, and temporal resources the correction system has available.

The rate is not incidental. It is the thing that determines whether the correction system will actually produce correction, because whether it produces correction depends on whether its throughput matches the throughput of the activity it is correcting. When the correction system's throughput is at least as high as the activity's throughput, the correction system catches the errors the activity produces before the errors accumulate into consequences the broader system cannot absorb. When the correction system's throughput is lower than the activity's throughput, the errors accumulate faster than the correction system can catch them. The differential between the two throughputs determines the speed at which unaddressed errors pile up in the system.

This is a quantitative claim about what looks like a qualitative phenomenon. We tend to think about correction systems in terms of their structure, the kinds of checks they perform, the authority they operate under, the thoroughness of the examinations they conduct. These features matter, but they do not determine whether the correction system will function at scale. What determines that is the ratio between the correction system's throughput and the activity's throughput. A meticulous correction system that processes a hundred cases a year cannot correct an activity that generates a million cases a year, no matter how excellent its procedures. It can only correct a sample, and the sample will be too small to prevent the errors that its processes would otherwise catch from propagating through the rest of the system.

The amplifiers this book has examined all have the same scale signature. They increase the throughput of the activity they amplify by orders of magnitude. The correction systems they displace have throughputs set by older parameters, which is to say, by the cognitive and institutional resources that produced the correction systems in the first place. The mismatch between the amplified throughput and the legacy throughput is where the collapse happens.

Consider the indulgence case in these terms. The pre-print practice of indulgence sale operated at a throughput determined by the number of clergy who could hear confession, assess penitents, and issue certificates by hand. The activity and the correction were structurally coupled: the same act that produced the indulgence also performed the assessment, because the assessment was the act of speaking with the penitent and evaluating the penitent's spiritual condition. The printing press decoupled them. It produced certificates at industrial rates, far exceeding the rate at which any actual assessment could occur, and the result was that the assessment ceased to happen in any meaningful sense. The certificate became a product, sold on the basis of payment rather than on the basis of the evaluative process that had formerly given it meaning. The correction system was not abolished. The Church retained its doctrinal authority, its institutional procedures, its hierarchy of review. What changed was that the correction system's throughput remained fixed at the pre-print level while the activity's throughput increased by a factor that cannot be precisely calculated but was certainly at least two orders of magnitude. The ratio became impossible, and the correction system collapsed in the specific sense that it ceased to perform correction on the activity it was nominally responsible for.

The harvest reporting case is the same architecture. Pre-amplification, a county magistrate could verify a harvest report by riding to the fields, speaking with farmers, observing the crop. The activity and the correction were again structurally coupled: reporting and verification used the same informational resources, because the report was drawn from the same local knowledge that a verifier would gather on visiting the location. The totalitarian reporting chain decoupled them. Reports moved up through the hierarchy at a rate determined by the speed of the bureaucratic apparatus, which was fast, while verification could only occur at the rate at which senior officials could physically visit local territories, which was slow. The correction system again retained all its formal features: the Party organs had the authority to investigate, the internal review procedures had not been dismantled, the apparatus of inspection existed. But the ratio between the reporting throughput and the verification throughput made the apparatus non-functional. By the time a report could be investigated, hundreds of subsequent reports had built on its premises. The correction system was not corrupted by any single decision. It was swamped by the volume of decisions that required correction.

The securitization case runs the same mechanism through a financial medium. A loan that stayed on a bank's balance sheet was corrected by the bank's own experience: if the borrower defaulted, the bank absorbed the loss, and the loss forced the bank to adjust its future underwriting. The correction throughput was set by the same mechanism that set the origination throughput, because both were activities of the same institution. Securitization separated origination from default absorption. Originators could produce loans at whatever rate the mortgage-broker pipeline allowed, which turned out to be very high. Default absorption was distributed across tranches of investors whose connection to the underwriting process was mediated by rating agencies, due-diligence firms, and the legal documentation of the securitization itself. The correction system that had existed at the origination stage was still present in attenuated form: underwriters still approved loans, compliance staff still reviewed documents, regulators still examined institutions. But the throughput at which the correction could actually track the origination activity was far lower than the throughput at which the activity was occurring, and the accumulated differential was the set of non-performing loans that eventually took down the system.

The Soviet case is the same mechanism running at the slowest time scale. A market price adjusts continuously, in real time, against the actions of all participants who take decisions on the basis of the price. The throughput of the price correction is effectively infinite at the scales relevant to most economic decisions, which is why market economies can produce coordinated behavior across very large populations without any central authority. Central planning replaced the price system with a reporting system whose throughput was bounded by the human and institutional capacity of the planning apparatus. The planners could update their plans on annual or quinquennial cycles. The economy they were trying to coordinate was shifting continuously, in response to weather, technology, consumer preferences, labor movements, and all the other factors that market prices integrate in real time. The ratio between the economy's actual change throughput and the plan's revision throughput was an enormous number, and the accumulated differential over seventy years was the stagnation that the system eventually could not absorb.

The architectural principle that these four cases make visible is that every amplification system has a native throughput, and every correction system has a native throughput, and the collapse is a function of the ratio between them. The collapse is not caused by the amplification per se. Amplification is not a problem in the abstract. It becomes a problem when the correction system that was calibrated to the pre-amplification throughput fails to scale with the amplified throughput. The correction system either scales to match, in which case the amplification produces the benefits it was built to produce without the accumulated errors the earlier throughput could not have produced, or it fails to scale, in which case the differential accumulates until the costs of the accumulated errors exceed whatever benefits the amplification was delivering.

Why Ambient Social Feedback Cannot Scale to Agent Throughput

What is specific about the agent economy is that the correction systems that humans have built for human cognitive failure are almost entirely ambient. By ambient I mean that they are features of the social environment rather than explicit mechanisms. The village priest, the skeptical colleague, the customer who rejects a bad offer, the spouse who pushes back on a half-formed plan: these are all correction systems, but they are not institutional artifacts built for the purpose of correction. They are features of the way humans live in communities, and their corrective function is incidental to their other functions.

Ambient correction systems have properties that institutional correction systems do not. They operate in parallel, because the social environment is many-to-many rather than one-to-many. They operate in real time, because they are built into the same interactions that produce the activity being corrected. They operate without dedicated capacity, because the corrector is doing something else as a primary activity and contributing correction as a byproduct. And they are calibrated to the throughput of the activity, because the throughput of the activity is set by the same human and social constraints that set the throughput of the ambient corrector.

These properties are what made ambient correction systems effective for the scale of human cognition as it operated before the amplifiers this book has examined. A peasant buying an indulgence could be corrected by an extended conversation with a village priest who knew the peasant's life and history. The conversation happened in the same interaction that produced the spiritual transaction, and the priest had no separate capacity to allocate to correction because correction was inseparable from the spiritual function he was performing. A village cadre reporting harvest figures could be corrected by the awareness that his neighbors were the ones doing the actual farming and would notice a lie. The correction was ambient in the social fabric, operating continuously without any formal mechanism. A bank officer making a loan could be corrected by the knowledge that the bank's capital was at stake and that the officer's own career was tied to the bank's survival. The correction was ambient in the institutional structure that the officer was embedded in. A producer in a market economy could be corrected by the price signal that emerged from the aggregate behavior of buyers and sellers. The correction was ambient in the market mechanism, operating at the full throughput of the economy's activity.

When the amplifiers stripped out the ambient correction systems, what replaced them was institutional correction: dedicated mechanisms with explicit procedures, specialized personnel, formal authority, and scheduled operation. The institutional correction systems were not inferior to the ambient ones in any simple sense. In many respects they were more rigorous, more explicit, more defensible against challenge. What they lacked was the throughput property that ambient systems automatically possessed. Institutional correction systems have dedicated capacity, which means they can do only as much as their dedicated capacity allows. When the activity they are correcting scales beyond their capacity, they cannot scale with it, because their scaling is constrained by the resources the institution is willing and able to allocate to the correction function.

The agent economy's correction problem is this problem in its most extreme form. The ambient correction systems that have historically contained human cognitive failure operate at human throughput: the throughput of conversation, of reflection, of social feedback, of market response. These throughputs are slow compared to agent throughput, which is measured in requests per second, workflows per minute, cross-agent handoffs per hour. An ambient correction system cannot operate at this throughput because ambient correction requires that a human being attend to the activity being corrected, and human attention has a throughput that is bounded by the same biological and social constraints that have always bounded it. The more agents are deployed, the more activity is occurring outside the reach of ambient correction.

The obvious response is to build institutional correction systems for the agent economy. Audit agents. Monitoring agents. Evaluation agents. There is considerable work happening along these lines, and some of it is useful. But the problem with institutional correction systems is that they are subject to the same scale constraints as any other correction system: their throughput is set by the resources the institution is willing to dedicate to them, and their throughput must match the throughput of the activity they are correcting if they are to actually perform correction. An audit agent that runs at one percent of the throughput of the agents it is auditing performs sampling correction, not comprehensive correction, and sampling correction catches only the errors that happen to fall into the sample. The errors that fall outside the sample propagate through the system unchallenged.

It is in principle possible to build correction systems that operate at agent throughput. It is not in principle possible to rely on ambient human correction to perform this function, because ambient human correction operates at human throughput, and the gap between human throughput and agent throughput is now large enough that ambient correction can cover only a trivial fraction of the activity. This is the specific problem that the agent economy is running into. The legacy correction systems that have contained cognitive failure historically are mostly ambient and operate at human throughput. The agent economy operates at agent throughput. The ratio between the two throughputs is already enormous and is growing. Every month, more activity is occurring at agent throughput with correction operating at human throughput, which means more activity is occurring without correction.

The Specific Shape of Corrective Systems That Could Work

The question is not whether correction systems can operate at agent throughput. They can. The question is what such correction systems look like, and what distinguishes the ones that actually perform correction from the ones that perform the appearance of correction without the substance.

The Risk Efficacy framework, introduced in Chapter 2 and deployed across the historical chapters, gives us the conceptual vocabulary for specifying what a corrective system has to do. A correction system has to perform four functions, corresponding to the four dimensions of Risk Efficacy: it has to assess the Calibration of the premises underneath the activity it is correcting, it has to test the Navigation logic that links premises to decisions, it has to monitor the Resilience of the updating loops that keep the activity in contact with reality, and it has to measure the Outcome Achievement of the activity against the outcomes it was deployed to produce. A correction system that does all four, at the throughput of the activity, is a functioning correction system. A correction system that does some of the four, at a fraction of the throughput, is a partial correction system that will fail in characteristic ways depending on which functions it handles and which it leaves unaddressed.

What would such a system look like for the agent economy? It would operate at agent throughput, which means it would itself be a computational system rather than an ambient human one. It would be embedded in the workflow that produces the activity, rather than being a separate review layer, because a separate review layer would have its own throughput constraints. It would operate on the premises that initiate queries rather than only on the outputs that conclude them, because the historical pattern shows that the failure mode enters the system through uncorrected premises and compounds through subsequent layers. It would carry information about outcomes back to the premises, in a feedback loop that updates the system's implicit calibration of which premises are likely to hold.

This description specifies a correction layer that does not currently exist in most agent deployments. Most agent deployments have a stack of functions that produce outputs and an evaluation apparatus that measures whether the outputs meet specified criteria. The evaluation apparatus is a correction system in the formal sense, but it is doing one of the four Risk Efficacy functions, Outcome measurement, and doing it at a throughput that is typically far below the activity throughput, and measuring against criteria that do not include the Calibration of the premises that initiated the query. An agent that takes a miscalibrated premise and produces an output that meets all the specified evaluation criteria is a pass from the perspective of most current evaluation apparatuses. The Calibration failure is invisible to the apparatus, because the apparatus is not designed to detect it.

The scale signature of the problem is legible. The architectural principle is specifiable. The collapse is not coming because correction is impossible. It is coming because the correction systems currently in place have throughputs calibrated to a pre-agent world, and the new throughput has arrived without the new correction.

A Note on Time Scales

One feature of the four historical cases deserves explicit treatment before the argument moves forward. The time scales across which the collapses unfolded vary by orders of magnitude. The indulgence collapse took a century to arrive. The Great Leap Forward collapsed within three years. The 2008 financial crisis accumulated over a decade and resolved in eighteen months. The Soviet collapse took seventy years. These are not the same duration, and the differences matter for what readers should expect from the agent economy.

The variation in time scales is not random. Each time scale is a function of three factors: the throughput differential between the amplifier and the correction system, the tolerance of the broader system for accumulated error, and the nature of the eventual corrective event. The indulgence case had a long arc because the broader system (European Christianity in the fifteenth and sixteenth centuries) had enormous tolerance for accumulated theological contradictions, and the corrective event (the Reformation) was a social and political upheaval that required a confluence of specific conditions to occur. The Great Leap Forward had a short arc because the broader system (peasant households dependent on the harvest) had zero tolerance for food shortage, and the corrective event was the biological reality of starvation, which arrives without any mediation. The 2008 crisis had a moderate arc because the broader system (the global financial architecture) had moderate tolerance, backed by central bank capacity, but the corrective event was the sudden unwillingness of wholesale funding markets to accept mortgage-backed securities as collateral, which propagated at the speed of telecommunications. The Soviet collapse had a very long arc because the broader system was politically insulated from short-term disciplining, but the corrective event, when it finally arrived, was a political collapse that had been delayed by decades of accumulated slack.

The agent economy's time scale is not knowable in advance with any precision, but the relevant determinants are visible. The throughput differential is very large, because agent activity is happening at machine speed and correction is happening at human speed. The tolerance of the broader system for accumulated error is moderate to high in most domains, because the errors are currently distributed across enough decisions that no single error produces catastrophic consequences. The nature of the eventual corrective event is not yet determined, but the historical pattern suggests it will take the form of a specific visible failure that makes the pattern undeniable: a major enterprise scandal, a regulatory action, a legal judgment, a publicized casualty count. The time scale will be determined by how long the distributed errors take to accumulate into an undeniable event, and by how long the post-event response takes to crystallize into structural change.

My estimate, which is explicit as an estimate and could be wrong, is that the first undeniable event will arrive within two to four years of the time this book appears, that the subsequent period of public pattern recognition will take one to three years after that, and that the structural correction will begin to emerge over a five to ten year horizon after the pattern becomes public. These estimates are bounded by the fact that the agent economy is advancing much faster than any of the historical cases, which compresses the accumulation period, while the institutional response is constrained by the same institutional dynamics that shaped the response to previous amplifier collapses, which may not compress as much. The compression differential is part of what makes the agent economy's correction problem distinct from the historical cases. The activity is accelerating toward a horizon at which the institutional response will still be moving at its historical pace.

The compression matters strategically. The window in which organizations can build correction layers in advance of a collapse is the window before the first undeniable event. After the undeniable event arrives, the correction becomes a reactive problem, and the costs of reactive correction are historically much higher than the costs of preemptive correction. The time scale is compressing. The differential is growing. Building correction infrastructure is not an optional improvement. It is a structural necessity on a horizon that current decision-makers can see.

Chapter 8: Uncalibrated Premises

The argument of the previous chapters has been architectural. It has described the shape of the failure mode, traced the shape through four historical cases, and named the principle that the shape embodies. This chapter brings the argument to ground. It walks through five enterprise domains in which the agent economy is already making, or is about to make, decisions on premises that no part of the agent stack has evaluated. For each domain I describe the specific cognitive failure that the agent will amplify, the specific correction system that will be outpaced, and the specific shape the collapse will take.

The selection of five domains is not exhaustive. The failure mode operates everywhere agents are being deployed to make decisions. The five domains I have chosen are the ones where the decisions are consequential enough that the failures will be visible, where the deployment timelines are short enough that the failures will arrive on readable horizons, and where the correction systems currently in place are legible enough that I can describe what will fail and how.

Before working through the domains individually, I want to state the general structure of what I am claiming, because the generality is easier to see once than once per domain. In each domain, the existing decision workflow has been produced by generations of adjustment between activity and correction. The decision itself takes place in a specific institutional context, within a specific regulatory environment, under specific accountability structures, with specific expert roles whose judgment is calibrated by training and experience. When an agent is inserted into the workflow to increase throughput, the agent produces decisions at a rate that the surrounding correction infrastructure was not built to absorb. The decisions may be on average better, worse, or about the same as the decisions the pre-agent workflow would have produced. What changes is that the rate overwhelms the correction system, which means errors that the correction system would have caught at lower throughput now pass through it. The errors accumulate, and at some point the accumulation produces a visible failure that the institutional apparatus has to respond to. The response is almost always slow, because institutional responses are slow.

The five domains are hiring, resource allocation, clinical triage, legal strategy, and operational planning. In each, I will spend some time on the specifics of the workflow before agent insertion, because the specifics are what determine the specific failure mode that agent insertion produces. Readers who are expert in a domain will find parts of the description familiar. The purpose of spending time on what they already know is to make visible the correction systems that the expert workflow depends on, because the correction systems are typically ambient and are often invisible to the people whose work they are correcting.

Hiring

The hiring decision is an extended negotiation between a candidate and an organization, conducted through a set of interlocking processes that have been adjusted over decades to balance several competing pressures. The organization needs to identify candidates who can perform the role, will remain in the role for some reasonable period, will fit with the organization's team dynamics, and will justify the compensation the organization is offering. The candidate needs to find a role where the work matches their capacities, the compensation meets their needs, and the organization's trajectory aligns with their career. The negotiation runs through resume review, application screening, recruiter conversation, hiring manager interview, team interview, reference check, offer negotiation, and the period between offer and start during which either party can still decide to exit.

At each stage of this workflow, correction systems are operating. The recruiter who has made a hundred placements has a calibrated sense of which candidates on paper actually perform in the role. The hiring manager who has managed for years has a calibrated sense of which interview behaviors correlate with on-the-job effectiveness. The team that conducts the later interviews has a calibrated sense of which candidates they can work with. References, when they are checked carefully, provide correction on what the candidate has actually done rather than what the candidate has said. The pre-start period allows both parties to continue evaluating the fit and to exit if new information emerges. The workflow has inefficiencies and produces bad hires at rates that vary by industry and role, but the workflow does, at volume, distinguish better from worse candidates with better than random accuracy, and it does this because the correction systems along the way actually correct.

Agent deployment in hiring is already extensive and is growing rapidly. The agent use cases include resume screening, candidate outreach, initial interviews, assessment scoring, and candidate ranking. The pitch is that the agent can handle far more candidates than a human recruiter can handle, which means the organization can cast a wider net, process a larger funnel, and surface candidates that would otherwise have been filtered out by the human bandwidth constraint. These claims are true at the level of throughput. The question is whether the increased throughput produces better decisions or more decisions of the same quality or worse decisions that are harder to detect because they come dressed in the legitimacy of a computational process.

The failure mode that agent hiring produces is a Calibration failure at the premise level. The premise that every candidate assessment rests on is a model of what the role actually requires, what the organization actually rewards, what the manager actually needs, and what the environment in which the work will be performed actually looks like. These premises are always imperfect in human hiring, but they are continuously updated through the accumulated experience of the humans involved. A recruiter who notices that the candidates who test well on a particular assessment do not actually perform well on the job will discount the assessment in future decisions. A hiring manager who notices that certain interview patterns produce hires who leave within a year will adjust the interviews. The feedback from performance to premise is noisy and slow, but it is present, and it keeps the premises in some approximate contact with reality.

Agent hiring workflows typically do not carry this feedback. The assessment agent produces rankings based on features that were correlated with positive hiring outcomes in a training set. The training set encodes a specific model of what positive hiring outcomes look like, and the specific model may or may not match what positive hiring outcomes look like in the organization deploying the agent. When the agent ranks candidates, the ranking is a function of the training set's model, not the deploying organization's actual performance data, and the gap between the two models is exactly the Calibration failure that the chapter's argument has been naming. The gap is invisible to the agent, because the agent has no access to the deploying organization's performance data in a form that would allow it to update its premises. The gap is invisible to the deploying organization, because the organization sees a ranking that presents itself as calibrated and sees no mechanism by which the ranking could be known to be miscalibrated.

The correction system that this setup outpaces is the ambient correction that was happening in the human workflow. A human recruiter who makes a bad call gets feedback from the hiring manager, from the team, eventually from the performance of the hired employee. The feedback shapes the recruiter's future judgments. An agent that makes the same call gets no equivalent feedback, because the agent's training does not update on the specific performance of the candidates it selected. The agent produces the next cohort of rankings using the same model it used for the last cohort, regardless of how the last cohort performed. The institutional correction, such as it is, happens through annual model retraining or through manual rule adjustments that the human operators of the agent implement. This is a correction system with a throughput of approximately one update per year, operating on an activity whose throughput is hundreds or thousands of decisions per year. The ratio is bad and is getting worse.

The shape of the collapse will be recognizable from the social science on hiring bias that has accumulated over decades. Organizations will hire cohorts of employees whose actual performance diverges systematically from what the ranking predicted. The divergences will be distributed non-randomly along the features the model implicitly weighted, which means the divergences will cluster by demographic group, by educational background, by the specific features that correlated with positive outcomes in the training set but do not correlate with positive outcomes in the deploying organization's actual environment. The collapse will become visible through performance data that the organization eventually connects to hiring data, through legal actions on behalf of candidates who were filtered out on grounds that turn out to be pretextual, and through the quieter failure of the organization to find and hire the people whose value would have been recognizable to a more calibrated process.

The institutional response will come through employment law, which already provides some of the apparatus. Disparate impact analysis, which already exists in US employment law, will be extended from its current application to cover agent-produced hiring decisions. The extension will require the legal system to develop competence in evaluating the internal architecture of hiring agents, which will take years. In the interim, organizations that have built agent-heavy hiring workflows will face a cumulative liability that grows with each cohort they have hired on miscalibrated premises. Some organizations will survive this liability through settlements and internal reforms. Some will not.

Resource Allocation

The resource allocation decision inside an organization is the decision about where to direct capital, headcount, attention, and time. It is made continuously, at many levels, and it shapes every other decision the organization makes. At the most senior level, it is the portfolio decision: which businesses to grow, which to maintain, which to divest. At the middle level, it is the project decision: which initiatives to fund, which to delay, which to kill. At the operating level, it is the task decision: which work gets done this week, which gets deferred, which gets dropped.

The workflow that produces these decisions is extensively developed in most organizations and is an object of consulting practice, financial analysis, and strategic theory. The workflow involves proposal generation, business case development, comparative review, prioritization, allocation, execution, measurement, and periodic reallocation based on observed results. The correction systems along the way include the experience of the senior decision-makers, the accountability of the middle managers whose recommendations will be tracked against outcomes, the scrutiny of finance functions that translate proposals into specific resource commitments, and the feedback from operations that eventually reveals which allocated resources produced the expected outcomes.

Agent deployment in resource allocation is at an earlier stage than in hiring but is developing rapidly. The pitches include the analysis of large datasets of business performance to identify allocation opportunities, the evaluation of investment proposals against historical returns from similar proposals, the simulation of allocation scenarios to identify the ones most likely to produce target outcomes, and the ongoing monitoring of allocated resources to identify opportunities for reallocation. The agent use cases span from individual analyst assistance to semi-autonomous recommendation systems that produce allocation suggestions for human review.

The failure mode that agent resource allocation will produce is an Achievement failure. The achievement dimension of Risk Efficacy measures whether the outputs of a decision process can be traded for outcomes of the kind the decision was meant to produce. In resource allocation, the question is whether the resources the system allocates produce the outcomes the organization deployed the resources to produce. The conversion from allocation to outcome depends on factors that extend well beyond what the allocation apparatus can observe: market conditions, execution quality, competitive responses, emergent customer preferences, macroeconomic shifts. A human allocation apparatus produces decisions that can be wrong about any of these factors, but the decisions are made by people who bear at least partial responsibility for the outcomes and who therefore build up, over time, a sense of which factors have been reliably predictive in their specific context.

Agent allocation systems will produce decisions that optimize against the predictors that were predictive in the training data. The predictors may be predictive in the deploying organization's context, or they may not be. The question cannot be answered in advance, because the answer depends on whether the deploying organization's context matches the context of the training data, and the match can only be evaluated by deploying the agent and observing the outcomes. The observation is slow, because resource allocation outcomes take months to years to fully resolve. During the observation period, the agent continues to produce allocation decisions, which means the agent produces decisions faster than the system can generate evidence about whether the decisions are correct.

The correction system that this outpaces is the accountability structure of the human allocation workflow. When a human analyst recommends an investment that fails, the analyst's future recommendations receive more skepticism, and the analyst's career progression reflects the failure. The accountability loop is imperfect, but it produces a cohort of senior allocation decision-makers whose judgments have been filtered by observed performance. An agent produces allocation recommendations without a personal accountability loop, and the institutional accountability loop (through model retraining, audit reviews, performance assessments of the agent deployment) operates at a throughput far below the throughput at which the agent produces recommendations. The ratio again becomes impossible, and the accumulated allocation decisions that were made on miscalibrated premises sit in the organization's portfolio producing results that nobody can cleanly attribute.

The shape of the collapse here will be less sharp than in hiring, because resource allocation outcomes are more diffuse and harder to track. It will appear as a gradual underperformance of organizations that have deployed agent allocation heavily relative to organizations that have maintained human judgment at the allocation layer. The underperformance will be attributable to many factors, which will make it difficult to isolate the agent contribution, but the differential will be observable in aggregate data that eventually someone will analyze. By the time the analysis is published, the organizations that have allocated heavily through agent systems will have embedded the allocation patterns in their portfolios in ways that are expensive to reverse. The institutional response will take the form of professional standards bodies for AI-assisted financial analysis, regulatory guidance on the use of AI in fiduciary decisions, and eventually litigation when institutional investors sue over allocation decisions that can be shown to have been made on systematically miscalibrated premises.

Clinical Triage

Clinical triage is the decision about what level of care a patient needs, how urgently, and from what kind of provider. It happens at every point of contact between a patient and the healthcare system, from the emergency room to the primary care call center to the pharmacy counter. The triage decision is consequential because it determines how much of the healthcare system's limited capacity gets allocated to the patient and how much delay the patient experiences before receiving appropriate care.

The workflow of clinical triage has been developed over decades and varies across healthcare systems in ways that reflect different structural approaches to the underlying problem. In a US emergency department, a nurse applies a standardized triage scale (Emergency Severity Index, for instance) to categorize arriving patients into acuity levels. In a primary care setting, a nurse or medical assistant screens incoming calls to determine whether the patient needs an appointment today, a same-day appointment, a routine appointment, or direction to an urgent care or emergency facility. The correction systems along the way include the clinical experience of the triage personnel, the escalation procedures when the triage is uncertain, the feedback from the downstream clinicians who receive the patient and can adjust the acuity rating, and the tracking systems that identify cases in which triage failures led to adverse outcomes.

Agent deployment in clinical triage is advancing on multiple fronts. Chatbot triage systems interact with patients to gather symptoms and produce a recommended level of care. Agent systems review patient records to identify cases that should be escalated to immediate clinical attention. Agent systems apply clinical decision support rules to the incoming flow of patients and produce priority orderings for the clinical staff. The throughput advantage of agent triage is real: a healthcare system can screen far more patients with an agent than with human triage staff, which expands access and reduces wait times for the cases that actually need immediate attention.

The failure mode here is potentially the most severe of the five domains, because the time constants are shortest and the consequences are biological. The Calibration failure in clinical triage takes the form of an agent that has been trained on a patient population whose base rates of specific conditions do not match the base rates in the population being triaged. An agent trained on urban emergency room data and deployed in a rural primary care setting will misjudge the probability that a given symptom profile indicates a serious condition, because the background incidence of serious conditions differs between the two settings. An agent trained on general medical data and deployed in a specialty oncology setting will misjudge symptoms whose specific meanings differ in the specialty context. An agent trained on historical patient populations and deployed as those populations shift, for instance, with the emergence of new infectious diseases or the changing demographics of an aging patient base, will misjudge cases whose statistical profile has changed since the training data.

These are not hypothetical failures. They are the expected result of deploying a statistical model trained on one distribution into an environment that presents a different distribution. In human triage, the mismatch is caught through clinical experience, because the humans performing the triage have a sense of what their specific patient population looks like and can override a textbook rule when the patient in front of them does not fit the textbook. In agent triage, the override mechanism depends on whether the human operator is reviewing the agent's recommendations and is present to override them, which depends on the throughput at which the agent is operating and the throughput at which the human operator can actually review.

The correction system that agent triage outpaces is the same ambient correction that all the previous examples have involved, but in the clinical setting the ambient correction has a much shorter time constant. A patient who was miscategorized returns to the system, either because their condition worsened or because they escalated their care seeking. The return event provides feedback. In a human triage workflow operating at moderate throughput, the returns are visible to the same triage personnel who made the original decisions, and the feedback adjusts their subsequent judgments. In an agent triage workflow operating at higher throughput, the returns are distributed across many subsequent triage events, and the feedback loop that would connect the returns to the original decisions has to be built explicitly in the system's architecture. Most current deployments have not built this loop, or have built it in a rudimentary form that does not actually update the agent's decision premises at the rate at which the returns are occurring.

The shape of the collapse in clinical triage will likely appear first as a cluster of adverse events in a specific healthcare system or a specific condition category. The events will be investigated. The investigation will identify that the triage decisions leading to the adverse events shared features that the agent's model had not weighted appropriately. The healthcare system will pull back the agent from the specific use case. The story will attract media attention. The regulatory response will follow, in the form of FDA guidance on clinical decision support, state medical board rulings on the supervision of AI clinical tools, and eventually insurance coverage decisions that require specific validation protocols for agent triage deployments. The response will likely reduce the throughput of the agents substantially, which will reduce their utility, which will create pressure for better-validated deployments to re-enter the market under the new regulatory framework. The cycle will take three to five years from the first visible adverse event to the establishment of a stable regulatory framework, and during the cycle the agent triage deployments will oscillate between expansion and retrenchment as the regulatory clarity develops.

Legal Strategy

Legal strategy is the set of decisions a party makes in pursuing or defending a legal matter: what to argue, what to settle, what to take to trial, what to appeal, how to structure a transaction, how to negotiate a contract, how to respond to a regulatory inquiry. The workflow is extensively developed in law firms and legal departments, and it involves the coordinated work of partners, associates, paralegals, and increasingly, specialized consulting services. The correction systems along the way include the judgment of senior lawyers with years of practice, the peer review that happens through case reviews and partner discussions, the feedback from clients about the outcomes of the strategies, the feedback from judges and opposing counsel about the quality of the arguments, and the reputational accountability that shapes the long-term career of lawyers whose strategies produce systematically bad results.

Agent deployment in legal strategy is already widespread in specific subtasks: document review, case law research, contract analysis, due diligence. The deployment is expanding into areas that were formerly considered the exclusive province of senior judgment: strategic planning, settlement valuation, argument construction. The pitch is that the agent can process a volume of precedent and factual material that no human lawyer can match, which allows the legal team to develop strategies informed by a more comprehensive analysis than was previously feasible. The claim is partially true at the level of information synthesis. The question is whether the synthesis translates into better strategy or into more strategy of the same quality or into a proliferation of strategic options that the human lawyers cannot evaluate carefully enough to choose well among.

The Navigation failure that agent legal strategy will produce is the failure to evaluate whether the premises underneath a strategic recommendation hold in the specific matter at hand. Legal strategy is a domain in which context dominates. A precedent that is controlling in one jurisdiction is persuasive in another and irrelevant in a third. An argument that succeeded against one kind of opposing counsel will fail against another. A settlement valuation that holds in one case structure is wrong in a structurally similar but legally distinct case. The human judgment that determines whether a strategic recommendation is good in the specific matter is a function of the senior lawyer's accumulated experience with the specific jurisdiction, the specific opposing parties, the specific judges, the specific industry context. The agent does not have this experience, and the agent's training data does not encode it in the form that would allow the agent to reproduce the senior judgment.

What the agent can do is produce recommendations that are plausibly correct on average across the training distribution. The average recommendation will be better than a random recommendation and will be produced at a throughput that far exceeds human capacity. The recommendations will be wrong in specific ways in specific matters, and the wrongness will not be visible to the agent or to the legal team that is using the agent unless the legal team is maintaining enough senior judgment in the review loop to catch the wrongness before it propagates into actual strategic decisions. The economic pressure to reduce senior lawyer involvement in the review loop will be substantial, because senior lawyer time is expensive and the agent's output looks like reasonable work product. As the economic pressure shifts more of the strategic work onto the agent, the correction system that the senior judgment provides will be outpaced, and the accumulated errors will begin to produce client outcomes that diverge from what the strategies promised.

The shape of the collapse will appear through malpractice claims and through client defections. Clients who experienced a failed strategy will sue their lawyers and will discover that the strategy was developed primarily by an agent, which will become a factor in the malpractice analysis. The state bar associations will begin to issue ethics opinions on the use of agent tools in legal practice. The opinions will tighten over time as the failures accumulate. The insurance carriers that write professional liability policies for law firms will adjust their underwriting to reflect the changed risk profile of firms that rely heavily on agent tools, and the adjustments will transmit pricing signals back into the firms that will alter their deployment decisions. The cycle will produce, over five to seven years, a set of standards for the supervised use of agent tools in legal practice that will stabilize the regulatory environment. During the cycle, clients whose matters were handled in the unstable period will have varying outcomes, with the losers being invisible except through the malpractice cases that eventually surface.

Operational Planning

The last of the five domains is the most diffuse and also the most pervasive. Operational planning is the set of decisions that determine how an organization runs from day to day: staffing schedules, inventory levels, maintenance cycles, capacity planning, logistics routing, supply chain coordination. Every organization of meaningful size engages in operational planning, and the accumulated decisions of operational planning constitute most of the work that actually determines whether the organization delivers on its commitments.

The workflow of operational planning has been deeply shaped by decades of computer support. Planning software is ubiquitous. Spreadsheets have been the standard tool for decades. Specialized planning systems exist for specific industries and functions. The correction systems along the way include the experience of the operations managers, the accountability of the functional leaders whose numbers will be audited against plan, the feedback from the operating units that must actually execute the plans, and the financial systems that translate plan execution into measurable outcomes.

Agent deployment in operational planning is one of the most actively developing areas of enterprise AI. The use cases span from individual planner assistance to autonomous planning agents that generate end-to-end operational plans and adjust them in real time as conditions change. The agents are built to ingest data from the operational environment, synthesize it against planning models, and produce recommendations or directly execute changes. The agents can operate at a throughput and integration that the previous generation of planning software could not match. Organizations deploying them frequently report significant productivity gains in the planning function.

The failure mode that agent operational planning produces is a Resilience failure. The resilience dimension of Risk Efficacy measures whether the system can update its model against new information at the pace at which the environment is changing. Operational environments change continuously, in ways that are partly predictable (seasonal patterns, promotional cycles, known disruptions) and partly unpredictable (sudden demand shifts, unexpected supply failures, emerging logistical constraints). Human operational planners have been adjusting to these changes for generations, and the adjustments embed a tacit knowledge about what kinds of changes require which kinds of responses. The tacit knowledge is very hard to encode in a planning model explicitly, which is why it tends to live in the senior operations staff whose judgment is what the planning software supports rather than replaces.

Agent planning systems typically encode the explicit knowledge and do not encode the tacit knowledge, because the tacit knowledge is not visible in the training data that produces the agent. When the operational environment produces a change that the tacit knowledge would have handled, the agent handles it using the explicit patterns it has learned, and the handling is often wrong in ways that the operational environment punishes quickly. The correction system that would have caught the wrong handling is the judgment of the human planners, but the economic pressure of the agent deployment is usually to reduce the human planning presence, because the agent's throughput makes human review look slow and expensive. The ratio of agent actions to human review falls, and the errors that the human review would have caught accumulate in the operations.

The shape of the collapse in operational planning will be messier than the other domains, because operational failures are continuous and are typically attributed to many causes. The specific contribution of agent planning will become visible through the pattern of failures that cluster around specific kinds of operational decisions: the ones where tacit knowledge was most important and where the agent's explicit patterns most diverged from what the tacit knowledge would have produced. The failures will show up as stockouts during specific kinds of demand events, as scheduling failures during specific kinds of workforce disruptions, as supply chain breakdowns during specific kinds of logistical stress. Organizations will begin to recognize the pattern and will adjust, but the adjustment will be slow, because the alternative, rebuilding the human planning expertise that was let go, is expensive and takes years to rebuild.

What the Five Domains Share

The five domains are different enough that the specifics of their failure modes look different. The convergence is in the pattern beneath the specifics: an amplifier is being deployed into a workflow that was previously bounded by the throughput of human cognition, the amplifier increases the throughput by factors that make human correction impractical, and the correction function that previously was performed ambiently is not being rebuilt as an explicit correction layer at the new throughput. The specific errors that the amplifier produces are different across domains, because the domains have different substrates. The architectural failure is the same.

What makes the five domains a coherent group is that in each one, the deployment decision is being made now, the throughput expansion is happening now, and the correction infrastructure is not being built now. The window in which the pattern could be interrupted is the current window, because after the first major visible failure in each domain, the correction infrastructure will begin to be built reactively, and reactive construction is historically more expensive than preemptive construction by margins that are large enough to be decisive for the organizations that experience the reactive costs.

The argument this chapter has made is that the premise of the building is specifiable, domain by domain, because the failure modes the building would correct are specifiable, domain by domain. The question that remains is whether the specification translates into construction. The answer is visible in the historical record of how the previous amplifiers were corrected, and in the trajectory by which the current failures will surface.

Chapter 9: The Trajectory of Visibility

The following scenario is a composite, drawn from accounts that surfaced in post-crisis investigations and congressional testimony across multiple firms. In the winter of 2006, a structured products trader at a regional investment bank noticed that the documentation on a specific mortgage bond pool did not match what the pool was supposed to contain. The discrepancy was small. The loans inside the pool had been underwritten to standards that were looser than the offering document suggested, and the pool's expected default rate was therefore higher than the investors who had bought the bond had been led to believe. The trader flagged the discrepancy to his supervisor, who flagged it to the firm's risk committee, who recorded the concern in a memo that was filed in the firm's internal compliance system. Nothing further happened. The pool traded normally. The trader's flag was one of thousands of flags that were being raised across the industry in the final eighteen months before the collapse, each one localized to a specific pool or a specific issuer or a specific counterparty, each one registering in internal systems that were not connected to each other and that had no mechanism for recognizing that the accumulating flags constituted a pattern rather than a series of unrelated concerns.

The flags were the canaries. They were not hidden. They were not deniable. They were recorded contemporaneously in the internal systems of firms that had every incentive to take them seriously, because the firms were the ones that would bear the losses if the flags were correct. The reason the flags did not produce action was not that the information was unavailable. It was that the information was available at a granularity and distribution that did not match the granularity and distribution at which decisions were made. Nobody at any firm was in the position of reading every flag, because the volume was too high and the flags were spread across too many desks. The firms that tried to aggregate their flags typically did so through committees that met quarterly, on time horizons that could not match the rate at which the amplifier was distributing the failure mode into the market. By the time the aggregation produced a pattern, the pattern was no longer a pattern being observed. It was a collapse being experienced.

The historical record of every one of the four cases this book has examined includes this same phenomenon: a period of eighteen to thirty-six months before the collapse became publicly visible, during which the evidence that a collapse was coming was accumulating in contemporaneous internal records at the institutions that would be most affected. The evidence was not acted on. It was not acted on because the institutions were not built to act on evidence of this kind at this rate, and the correction mechanisms that could have acted on the evidence had been stripped out by the amplifier's architecture. The collapse, when it came, was described as a surprise. The surprise was not that the failure occurred. The surprise was that the failure became visible at the level at which decisions were being made, after a long period of being visible only at the level at which the decisions were being generated.

The agent economy is now in this window. The flags are beginning to accumulate in enterprise environments where agents have been deployed into decision workflows. The flags are being recorded in the internal systems of the deploying firms. The flags are spread across desks and functions and use cases in ways that mirror exactly the distribution pattern that preceded the previous collapses. The question this chapter takes up is what the accumulation of flags will look like over the next eighteen to twenty-four months, what the specific sequence of public visibility will be, and what kinds of events will make the pattern legible to observers who are not currently looking for it.

The Three-Event Pattern

Historically, the transition from invisible accumulation to visible collapse has followed a three-event pattern. The first event is an incident in a specific organization or market that looks like a one-off failure. The second event is a structurally similar incident in a different organization or market that begins to suggest the failure might be systemic. The third event is a cluster of incidents across multiple organizations that makes the systemic character of the failure undeniable. The pattern is visible in the 2008 collapse, where the failure of Bear Stearns in March 2008 looked like a one-off, the failure of Lehman Brothers in September looked like an industry event, and the subsequent distress across the entire investment banking sector made the systemic character visible. The pattern is visible in the Soviet collapse, where the 1989 events in Poland looked like a one-off, the 1989 events across Eastern Europe looked like a regional pattern, and the 1991 dissolution of the Soviet Union itself made the systemic character visible. The pattern is visible in the Reformation, compressed on a longer time scale, where the Wittenberg theses looked like a local dispute, the spread of the movement through German-speaking Europe looked like a regional dispute, and the Peace of Augsburg made the systemic character of the theological challenge visible.

What the pattern shows is that the collapse is not typically a single moment. It is a sequence of events that becomes visible as a pattern only after the third or fourth iteration. The first iteration is not recognized as the beginning of a collapse. It is described as an anomaly. The second iteration is not recognized as the second iteration. It is described as a different anomaly. The third iteration is when the observer community begins to construct the narrative that the first two iterations were connected, and by the third iteration the collapse is already well underway.

The agent economy collapse, if it follows the pattern, will be visible first as a specific incident in a specific enterprise. The incident will have the shape of an agent-deployed workflow producing an outcome that diverges substantially from what the workflow was supposed to produce, with material consequences for the organization or its customers. The incident will be reported in the trade press for the relevant industry. It will be described as a deployment error, a vendor failure, a specific bug in a specific model. The framing will suggest that the incident was avoidable and that other deployments are not subject to the same failure mode.

The second incident will arrive within six to twelve months. It will occur in a different industry, under a different vendor, using a different agent architecture. The incident will have the same structural shape: a workflow that produced a miscalibrated outcome with material consequences. The trade press coverage of the second incident will draw comparisons to the first but will typically frame the comparison as a general observation about AI risks rather than as an identification of the specific architectural failure mode that both incidents share. The deploying firms will respond with additional oversight procedures, additional vendor review processes, additional internal monitoring. The responses will not address the architectural failure mode, because the failure mode has not yet been recognized as architectural.

The third cluster of incidents will arrive in the twelve-to-eighteen-month window after the second incident, and it will be the cluster that makes the pattern visible. The cluster will include incidents in multiple industries, across multiple vendors, with structurally similar shapes. The cluster will be reported with language that begins to approximate the architectural description: observers will begin to speak about agent deployments as a category that produces a specific kind of failure, rather than as individual deployments that each happened to fail for individual reasons. The shift in language will be the signal that the invisible accumulation has become visible. From that point forward, the collapse will be experienced as a known problem rather than as a surprise, which will shift the political and regulatory dynamics significantly.

The Likely Shape of the First Event

The first event will most likely arrive in one of three domains, based on the analysis of the previous chapter: clinical triage, employment decisions, or financial services. These domains share two features that make them likely venues for the first visible failure. The first feature is that they have relatively short time constants between the decision and the observable outcome, which means that miscalibration can be detected in weeks or months rather than years. The second feature is that they have existing legal and regulatory frameworks that create formal channels for surfacing failure, which means that the incidents become public through mechanisms that compel disclosure rather than through mechanisms that depend on voluntary reporting.

If the first event arrives in clinical triage, it will likely take the form of a hospital system that discovers a cluster of adverse patient outcomes that trace to a triage decision made with material input from an agent system. The cluster may be small, perhaps a dozen cases over several months, but the legal reporting requirements around adverse medical events will force the cluster into public view. The hospital system will pull back the agent from the triage function, the state medical board will open an investigation, and the story will attract medical trade press coverage that frames the incident as a cautionary tale about AI in healthcare. The framing will identify the specific agent deployment as the problem and will not generally extrapolate to the architectural question of whether other agent deployments in clinical contexts are subject to the same failure mode.

If the first event arrives in employment, it will likely take the form of a class action lawsuit brought by job candidates who were filtered out of hiring processes at a specific large employer. The lawsuit will allege disparate impact on a protected class, supported by statistical evidence about the distribution of candidates the agent's filtering produced. The employer's defense will invoke the vendor's claims about the agent's calibration, which will put the vendor in the position of defending the agent's training data and architecture in a legal proceeding. The proceeding may produce a settlement before trial, in which case the settlement will be confidential but the facts of the allegation will have entered the public record through the complaint. Trade press coverage will emphasize the legal exposure of employers who deploy agent hiring tools, and human resources consulting firms will begin publishing advisories that recommend specific audit protocols for agent deployments. The framing will focus on legal risk mitigation rather than on the architectural question underneath.

If the first event arrives in financial services, it will likely take the form of a trading or allocation error at a mid-sized institutional investor that becomes visible through performance disclosure or regulatory filing. The error will involve an agent system that generated trades or allocations on premises that did not match the actual state of the market. The institutional investor will report the loss to its limited partners. The limited partners will demand explanation. The explanation will become the subject of litigation when the partners argue that the investor's duty of care was violated by delegating the decisions to an inadequately supervised agent. The litigation will surface the internal documentation of the agent's deployment, which will reveal that the internal flags about the agent's behavior had been present for months before the visible loss. Financial trade press will cover the litigation with a focus on fiduciary responsibility in the use of AI, and consulting firms will respond with governance frameworks for institutional AI adoption.

Each of these three scenarios produces the first event in a form that is publicly visible but that is framed in ways that do not yet identify the architectural failure mode. The first event, in any of these forms, will be described as a specific problem with a specific deployment, correctable through better procedures and better supervision. The description will be partially accurate, because better procedures and better supervision will improve the specific deployment that produced the first event. The description will be substantially incomplete, because the failure mode is not specific to the deployment. The failure mode is architectural.

The Lag Between the Event and the Pattern

The most important feature of the trajectory is the lag between the occurrence of the first event and the recognition of the pattern. The lag is not an accident. It is produced by specific features of how information about failures travels through professional and regulatory systems. Understanding the lag is critical, because the period of the lag is the period during which the failure mode continues to propagate while the observer community is still treating the early events as isolated.

The first feature that produces the lag is the siloing of trade press. Healthcare trade press does not read employment trade press. Employment trade press does not read financial trade press. Each industry has its own publications, its own conferences, its own communities of practice. When an incident occurs in one industry, the reporting is optimized for the readers of that industry's trade press, which means the framing is industry-specific. Observers who read across industries, if they exist at all, are rare, and they are typically not in positions to drive institutional response. The siloing means that an incident in healthcare does not inform the assessment of agent deployments in employment, even though the underlying failure mode is the same.

The second feature is the time horizon of consulting reports. The major consulting firms produce analyses of emerging risks on cycles of six to twelve months. An incident that occurs in March will appear in consulting reports that publish in the fall, if the consulting firm chooses to include it in the research agenda. If the first major event in the agent economy collapse occurs early in the predicted window, the first consulting report that puts it into a broader frame will likely appear nine to fifteen months later. The consulting report will typically frame the incident as a specific case of a broader category of risks, which begins the process of architectural recognition, but the framing will be cautious, because the consulting firm has relationships with agent vendors and with firms that are in the middle of large agent deployments. The cautious framing will slow the recognition even further.

The third feature is the speed at which regulatory agencies build institutional competence in new technical domains. The regulators who will eventually adjudicate the agent economy failures typically lack the internal capacity to evaluate the architectural features of agent deployments at the time the first incidents arrive. Building the capacity takes staff hires, training programs, and the development of analytical frameworks that can be applied consistently across cases. The process takes one to three years from the first regulatory hearing to the first regulatory framework that embodies an architectural understanding of the failure mode. During the building period, the regulatory response is reactive and case-specific, which addresses individual deployments but does not address the architectural pattern.

The fourth feature is the delay between litigation initiation and litigation resolution. Employment discrimination cases typically take two to four years from filing to judgment. Medical malpractice cases typically take three to five years. Securities fraud cases typically take three to seven years. The facts that surface through discovery in these cases are often transformative for the public understanding of what happened inside the defendant institutions, but the surfacing happens years after the events. During the years of the proceedings, the public information about the cases is limited to the initial complaint and the routine procedural filings, which means that the observer community has very little information to work with about what the defendants actually did.

The combination of these four features produces a lag of eighteen to thirty-six months between the first event and the general recognition of the architectural pattern. During the lag, agent deployments will continue to expand, flags will continue to accumulate in internal systems, and the invisible stage of the collapse will continue to develop. The lag is not a failure of attentiveness. It is the default operating speed of the institutions that are supposed to translate incidents into public understanding. The observer who tries to collapse the lag will be framed as alarmist, because the evidence that would justify the alarm has not yet been aggregated through the channels that the observer community recognizes as authoritative.

Canaries Worth Watching

A reader who wants to track the trajectory directly, without waiting for the observer community to reach recognition, can watch for specific canaries. The canaries are signals that become visible earlier in the trajectory than the major public incidents, and they indicate that the underlying failure mode is propagating at a rate the public coverage has not yet caught up to.

The first canary is the emergence of agent-specific audit and compliance services. When consulting firms begin selling agent audit services, and when the services begin to specify audit protocols that address calibration, navigation, resilience, and outcome achievement, the existence of the services is a signal that enterprise buyers have begun to sense the problem. The audit services are a lagging indicator of enterprise concern and a leading indicator of public disclosure. When a firm buys an agent audit and the audit surfaces problems, the problems get recorded in documentation that becomes discoverable in later litigation. Watching the audit service market is a way of watching the formation of the litigation pipeline.

The second canary is the behavior of enterprise AI insurance. The insurance carriers that write professional liability coverage for firms using agent tools are actively developing the actuarial models that will price the coverage. As the models mature, premiums for specific use cases will rise or fall in ways that reveal the carriers' private judgments about which deployments are most likely to produce claims.

The premium movement is a canary because insurance carriers have access to internal data on claims that predates public disclosure. Rising premiums in specific agent use cases, without corresponding public incidents, indicate that the carriers are seeing claim patterns in the data that the public has not yet seen.

The third canary is the behavior of specific senior executives at agent vendor firms. When a senior technical executive at a major vendor leaves on short notice, the departure may be routine, or it may indicate that the executive has decided the vendor's public claims about its products cannot be defended with confidence. The pattern of departures is a signal that only becomes legible in retrospect, but an observer who tracks the pattern contemporaneously can form tentative hypotheses about which vendors are facing internal doubts about their products' safety. The same signal is visible in the opposite direction: executives who join vendor firms carry signals about what those executives believe about the vendor's trajectory. The combination of incoming and outgoing senior technical staff at major vendors is a canary worth monitoring.

The fourth canary is the regulatory filings of publicly traded firms that have deployed agents at scale. The required disclosures about material risks include descriptions of AI-related risks that the firm believes could be material. The language of these disclosures is typically lawyered into vagueness, but the evolution of the language over time is informative. When firms begin describing agent-related risks in more specific terms, naming specific use cases and specific failure modes, the evolution signals that internal legal and risk functions have begun to judge the risks as concrete enough to warrant specific disclosure. The shift from vague to specific language is a canary that indicates the firm's internal documentation has moved in a direction that the firm's lawyers believe needs to be acknowledged publicly.

The fifth canary is the emergence of academic and industry conferences on agent governance. When the conferences begin to include sessions specifically on agent failure modes, calibration problems, and correction architectures, the sessions signal that the research community has begun to engage with the architectural question. The research community engagement is a lagging indicator of enterprise concern but a leading indicator of policy response, because the regulatory and legislative actors who will eventually act on the failure mode tend to draw their analytical frameworks from the research community's output. Watching the conference programs of the major AI governance venues is a way of watching the development of the policy vocabulary that will eventually be applied to the collapse.

The five canaries are not predictive in the sense of giving exact timing. They are indicators that the underlying process is progressing. An observer who tracks them across months will develop a sense of the rate at which the trajectory is moving, which is more useful than any single prediction of when the major events will occur. The rate, in turn, gives the observer a window in which to act, either to reduce their own exposure to the failure mode or to participate in the construction of the correction infrastructure that the failure mode will eventually demand.

What Happens When the Pattern Becomes Visible

When the three-event pattern completes and the architectural failure mode becomes legible to the broader observer community, the political and regulatory dynamics will shift substantially. The shift will have the shape that previous collapses have shown. The first phase will be a period of recrimination, during which the institutions that deployed agents at scale will be asked to account for why they did so despite the internal flags that were accumulating. The accounting will be partial and defensive, and much of it will focus on blame assignment to specific individuals or specific vendors rather than on the architectural question. The phase will last approximately twelve to eighteen months and will produce significant reputational damage to firms that are unable to demonstrate that their deployments included adequate correction mechanisms.

The second phase will be a period of regulatory construction, during which the relevant regulatory agencies will build the frameworks that will govern agent deployments going forward. The frameworks will likely incorporate some version of the four-dimensional vocabulary that this book proposes, either in those exact terms or in equivalent formulations that distinguish between premise calibration, workflow navigation, environmental resilience, and outcome validation. The frameworks will take three to five years to mature into stable regulatory regimes, and during the maturation period, agent deployment will be slower and more expensive than it has been in the pre-collapse phase.

The third phase will be a period of institutional reformation, during which firms and industries will rebuild their agent governance architectures to accommodate the new regulatory frameworks and to restore public trust. The rebuilding will produce a more capable ecosystem of agent deployment than existed before the collapse, because the rebuilding will have been informed by the architectural lessons of the failure. The rebuilding will also produce substantial winners and losers. The winners will be the firms that had already begun building correction infrastructure before the collapse forced the industry to do so. The losers will be the firms that had deployed most aggressively without correction infrastructure, and that will find themselves both reputationally damaged and operationally unable to adapt to the new regulatory requirements on a competitive timeline.

The entire trajectory, from the first visible event to the establishment of stable post-collapse institutions, will likely span five to eight years. Five years is a short time relative to the sweep of the previous collapses. Eight years is long enough that the readers of this book will have been living through the collapse during its development. The point of laying the trajectory out in advance is to give readers a frame for interpreting events as they occur, rather than forcing them to construct the frame after the fact. The frame is not a prediction of specific events. It is a structure against which specific events can be located, so that the first does not look like a one-off, the second does not look like a different one-off, and the third does not have to arrive before the pattern becomes recognizable.

What remains is to examine, across the four historical cases, what post-collapse institutional construction has actually looked like, and then to describe what preemptive construction could look like for the agent economy, if the window between now and the first visible event remains long enough to permit it.

Part IV: What Gets Built After

The first three parts of this book have described a failure mode, traced its historical recurrences, and projected its likely trajectory in the specific domain where it is now unfolding. Part IV takes up a different question. If the failure mode is architecturally predictable, and if the post-collapse response is also architecturally predictable, what does the predictable response actually look like? And if the predictable response is visible in advance, is there a way to build it in advance, rather than waiting for the collapse to force its construction?

The two chapters of Part IV address these questions in sequence. Chapter 10 examines the post-collapse institutional construction that followed each of the four historical cases, with particular attention to what the institutions were built to do and what they could not do. Chapter 11 describes the architecture of a calibration layer for the agent economy and makes the argument for preemptive rather than reactive construction.

The tone of Part IV shifts modestly from the preceding chapters. Parts I through III have been diagnostic. Part IV is constructive. The shift does not mean the diagnostic rigor loosens. It means that the diagnostic framework is now being applied to the question of what should be built, which is a different kind of analysis from the question of what has failed.

Chapter 10: Post-Collapse Construction

The peace that ended the German Wars of Religion was signed at Augsburg in September 1555, thirty-eight years after Luther had posted the theses at Wittenberg. The treaty's operative principle, cuius regio, eius religio, assigned the confessional identity of a territory to its ruler, which meant that the unified medieval Christendom was formally and explicitly replaced by a system in which confessional pluralism was embedded in the legal architecture of political sovereignty. The treaty was a defeat for every party that had entered the conflict hoping to win. The Catholic powers had not reunified Christendom under Rome. The Lutheran powers had not converted the Catholic territories. The reformed theologians who had hoped to rebuild the church on scriptural foundations had produced instead a permanent fracture. What the treaty accomplished was the creation of an institutional framework that could accommodate the fracture without requiring either side to resolve it on the other side's terms.

The framework was not a solution to the amplification problem that had produced the fracture. It was a response to the collapse the amplification had caused, designed to prevent further collapses of the same kind by accepting the new reality that the amplifier had created. The medieval architecture of unified religious authority was not restored. It could not be restored, because the printing press had made the informational substrate on which that authority had rested obsolete. What was built instead was a political architecture in which religious authority was partitioned into units small enough that the internal coherence of each unit could be maintained without requiring conformity across units. The Peace of Augsburg is the prototype of post-collapse institutional construction, and its features are legible across every subsequent case of the same kind.

What follows examines what was built after each of the four historical collapses, with attention to three questions. Which institutional features did the post-collapse construction introduce that had not existed before? Which dimensions of the failure mode did those features address, and which were left unaddressed? And what were the costs of building reactively, after the collapse, rather than preemptively, before it? The answers are the basis on which the next chapter specifies what a preemptive construction for the agent economy could look like.

The Modern State: What the Reformation Produced

The post-Reformation construction that eventually stabilized into the modern state was not a single act. It was a sequence of constructions that unfolded over roughly a century, from the Peace of Augsburg in 1555 to the Peace of Westphalia in 1648, with the subsequent consolidation of sovereign territorial authority extending across the following decades. The sequence produced several specific institutional features that had not previously existed in the medieval world, and each feature addressed a specific element of the failure that the printing press had amplified.

The first feature was the principle of territorial sovereignty itself. The medieval order had overlapping jurisdictions: ecclesiastical authority, imperial authority, royal authority, feudal authority, municipal authority, each claiming different scopes over different matters. The overlap had been workable when the informational flows that sustained authority were local and when claims of authority could be adjudicated face to face, through courts and councils that dealt with specific disputes case by case. The printing press broke the workability by allowing claims to propagate across the overlapping jurisdictions faster than the overlapping jurisdictions could coordinate a response. Territorial sovereignty addressed the problem by assigning each piece of territory to a single ultimate authority, which meant that the overlapping authority disputes of the medieval order were resolved in favor of whichever authority held the territory. The resolution was blunt but it was functional. It allowed the state to become the unit that could, in principle, monitor informational flows within its own territory and act on them in a coordinated way.

The second feature was the development of administrative bureaucracy capable of operating at the scale of the territorial state. Medieval administration had been conducted largely through personal retainers and ecclesiastical offices. The bureaucratic state that emerged after the Reformation built institutional offices that were not reducible to the individuals who held them, which meant that administrative capacity could be transmitted across generations and scaled up as the territory's needs expanded. The bureaucracy was the mechanism by which the sovereign could act consistently across the territory, and its construction took decades of institutional experimentation that produced, among other things, the first true standing civil services in early modern Europe.

The third feature was the state's assumption of authority over institutions that had previously been independent. Education is the clearest case. Before the Reformation, education in most of Europe had been a function of the church, conducted through ecclesiastical schools and universities that received their charters and their curricula from Rome. After the Reformation, the state began to assert authority over education within its territory, which allowed it to shape the content of what was taught and to monitor the informational flows that passed through educational institutions. The assumption of educational authority was not clean and it was not complete. Churches retained significant educational roles well into the modern period. But the principle that the state had legitimate concern about what its subjects learned was established in the post-Reformation settlement, and its establishment was a direct response to the informational amplification that the printing press had produced.

The fourth feature was the development of the press licensing and regulatory apparatus itself. Every post-Reformation state eventually developed some version of a licensing regime for printed materials, ranging from the formal censorship systems of Catholic territories to the copyright and libel regimes that emerged in Protestant territories. The regulatory apparatus was a direct response to the amplifier, and it was calibrated to the specific features of the amplifier that had produced the most severe effects: unlicensed reproduction, anonymous authorship, rapid distribution across jurisdictional lines. The apparatus was not effective at containing the amplifier, because the amplifier had already reshaped the informational environment in ways that the regulatory response could moderate but not reverse. What the apparatus did accomplish was the incorporation of informational regulation into the standing functions of the state, which meant that future informational amplifiers would arrive into an environment that already had regulatory infrastructure in place, even if the infrastructure would have to be adapted to each new amplifier's specific features.

What the modern state did not address was the underlying failure mode. The miscalibration of religious belief, which had been the original cognitive failure that the indulgence economy amplified, was not corrected by the modern state. It was partitioned. The state allowed different territories to maintain different miscalibrations and prevented the miscalibrations from coming into direct conflict with one another through the sovereignty framework. The partition was a containment strategy, not a correction strategy. It preserved the failure mode within each partition while preventing cross-partition amplification. Similar containment strategies will be visible in each of the other three post-collapse constructions, and their shared shape will be relevant for the argument of Chapter 11.

Dodd-Frank and Stress Testing: What 2008 Produced

The institutional construction that followed the 2008 financial crisis is the post-collapse case closest to us in time, and the one whose internal architecture is most visible in the contemporary documentary record. The construction was centered on the Dodd-Frank Wall Street Reform and Consumer Protection Act of 2010 and on the regulatory apparatus that the act created, including the Consumer Financial Protection Bureau, the Financial Stability Oversight Council, the Office of Financial Research, and the expanded regulatory authority granted to the Federal Reserve over systemically important financial institutions. The construction also included the Basel III international banking standards, which were developed in parallel and which introduced new capital and liquidity requirements for banks operating across national jurisdictions.

The post-2008 construction addressed several specific features of the failure that the securitization amplifier had produced. The first feature was the fragmentation of responsibility across the securitization chain. Dodd-Frank imposed specific duties on each link in the chain, including risk retention requirements that required originators of securitized loans to retain a portion of the risk on their own balance sheets. The requirement was designed to restore, at least partially, the feedback loop that had been broken when originators had been able to transfer all of the risk out of their balance sheets through securitization. The construction did not fully restore the feedback loop, because the retained portion was small and could be hedged, but the requirement did reinject some measure of downside exposure into the origination decision.

The second feature was the opacity of the securitized instruments. The crisis had revealed that the ratings agencies had not been able to assess the risk of the securitized tranches accurately, and that the buyers of the tranches had not been able to independently verify the ratings. The post-2008 construction addressed this through increased disclosure requirements, standardized reporting frameworks, and the expansion of regulatory authority to monitor the ratings agencies themselves. The construction did not produce complete transparency, because the underlying complexity of the instruments persisted, but it did produce substantially more information flowing into the regulatory apparatus and into sophisticated market participants.

The third feature was the absence of a system-level view of counterparty exposure. Before 2008, no entity had a comprehensive view of how the major financial institutions were exposed to each other through derivative contracts, securitized positions, and short-term funding arrangements. The failure of Lehman Brothers produced cascading losses across counterparties precisely because the counterparty structure had been opaque. The post-2008 construction addressed this through the creation of the Office of Financial Research and through the expansion of derivative reporting requirements that channeled contract data into central repositories. The construction made counterparty exposure visible to regulators in ways it had not been visible before, which allowed the regulators to identify concentrations of risk that might warrant intervention.

The fourth feature, and the most significant for the argument that follows, was the introduction of stress testing as a standing regulatory function. Stress testing requires systemically important financial institutions to model their balance sheets against specified adverse scenarios and to demonstrate that they could remain solvent through the scenarios. The stress testing function is, in the vocabulary used throughout these chapters, a form of calibration checking. It does not assume that the institutions' internal models are correctly calibrated to the risks they face. It imposes external scenarios against which the institutions' models must be tested, and it uses the discrepancy between the institutions' projections and the regulatory scenarios as a signal about where the internal models may be miscalibrated. The stress test is an explicit correction mechanism, built to operate at the scale and throughput of the post-securitization financial system, and designed to catch calibration failures before they produce systemic consequences.

Stress testing is the feature of the post-2008 construction most directly relevant to the agent economy problem because it operationalizes the kind of correction the argument here calls for. A stress test works by specifying a set of external conditions that the system under test must remain functional through, measuring the system's response to those conditions, and using the response to identify places where the system's internal model of itself diverges from its actual behavior under stress. The function is explicitly about calibration. It is explicitly about catching failures before they become visible. And it is explicitly designed to operate at the throughput of the system it monitors.

What the post-2008 construction did not do was eliminate the architectural features of securitization that had produced the crisis. Securitization persists. The market for mortgage-backed securities, asset-backed securities, and collateralized loan obligations is larger now than it was in 2007, although the specific features of the pre-crisis market have been modified. The construction was a containment strategy, not a reversal. It accepted that securitization would continue to be a feature of the financial system and built regulatory apparatus around it that is designed to catch the specific failures that securitization is prone to produce.

Market Infrastructure: What the Post-Soviet Transition Produced

The post-Soviet transition is the most heterogeneous of the four cases, because the specific path of reconstruction varied dramatically across the countries that emerged from the Soviet system. The most instructive case is probably Poland, which pursued the Balcerowicz shock therapy program in 1990 and which had, by the late 1990s, successfully rebuilt market infrastructure and achieved sustained growth. Russia's path was different, and more ambiguous, and produced an oligarchic economy whose features diverge significantly from Poland's. East Germany's path was different again, because its reconstruction was subsumed into the institutional framework of the Federal Republic through reunification. The Baltic states pursued yet another path, centered on rapid integration with Western European institutions. The variation is instructive, because it shows that the construction after collapse is not determined by the collapse alone. It is shaped by the specific institutional choices made in the post-collapse period.

What all of the successful reconstructions shared was the rebuilding of market infrastructure that had been destroyed or had never existed under the Soviet system. The core infrastructure includes: functional banking systems with independent risk assessment, property rights frameworks that allow assets to be owned and transferred, contract enforcement mechanisms that make commercial agreements reliable, currency systems with credible monetary policy, and commodity and securities markets where prices can form through the interaction of buyers and sellers. None of these existed in the Soviet system in forms that could function in a market economy. Their construction was the substantial work of the transition.

The construction was, in the vocabulary of this book, the restoration of feedback mechanisms that the central planning amplifier had eliminated. Prices in the post-transition economies could again carry information about scarcity. Profit and loss could again register whether enterprises were producing value or destroying it. Bankruptcy could again eliminate producers whose outputs were not worth their inputs. Each of these mechanisms is a correction system that operates at the throughput of the economy it governs, which is very high, because market decisions are made continuously and in parallel across millions of agents.

The infrastructure that was built in the successful transitions was, to an important degree, infrastructure that had been developed in Western market economies over the preceding two centuries. The post-Soviet states imported institutional templates from countries where the infrastructure already existed, adapted them to local conditions, and implemented them with varying degrees of success. The adaptation was not a simple technology transfer. It required building the administrative and legal capacity to operate the institutions, training the professional cadres who would staff them, and developing the social norms that would sustain them. The decade of the 1990s was substantially consumed by this work in the countries that pursued it seriously.

What the post-Soviet construction accomplished was the rebuilding of resilience, in the specific sense of the framework. The central planning amplifier had eliminated the mechanisms by which the economic system could update against local conditions. The post-Soviet infrastructure restored those mechanisms. The restoration was uneven across countries, and the unevenness is still visible in the economic performance of the various post-Soviet states three decades later. But the principle of the restoration is clear, and the institutional features that enabled it can be identified specifically. They are the features that allow a complex distributed system to maintain calibration against reality through the aggregation of many local decisions, rather than through the specification of a central plan.

What the post-Soviet construction did not do was produce perfect institutions. The oligarchic capture that occurred in Russia is the most visible failure mode, but similar patterns of imperfect institutional construction can be identified in other post-Soviet states. The imperfections are not random. They reflect the specific institutional choices made during the transition and the specific political economy of the transition period. The lesson is that post-collapse construction is shaped by the resources and choices available at the time of construction, and that different choices produce different qualities of institutions, even when the general pattern of what needs to be built is clear from the historical record.

Deng's Reforms: What the Great Leap Forward Produced, Thirty Years Later

The post-Great-Leap-Forward construction is the case with the longest lag between the collapse and the eventual institutional response. The famine ended in 1962. The Cultural Revolution, which partially functioned as a political suppression of the recognition of the

Great Leap Forward's failures, ran from 1966 to 1976. The substantive reform period did not begin until 1978, sixteen years after the famine and two years after Mao's death. Deng Xiaoping's reform program, which extended from the late 1970s through the 1990s, was the eventual institutional response to the failure that the Great Leap Forward had made visible, and the long lag between the collapse and the response is one of the most important features of the case.

The Deng reforms introduced several specific institutional features that had not existed in the Maoist system. The first was the household responsibility system in agriculture, which restored the connection between agricultural effort and agricultural reward that the communes had broken. Peasants who produced more could retain more, which produced immediate and dramatic increases in agricultural output. The feature was a restoration of Achievement in the specific sense of the framework. The system allowed agricultural effort to produce measurable outputs that could be converted into improvements in the household's circumstances, which is the basic feedback loop that the commune system had eliminated.

The second feature was the introduction of Special Economic Zones, which allowed market-based economic activity to develop in geographically bounded areas while the rest of the country continued to operate under the planned system. The zones were, in effect, a controlled experiment in market economics that produced results the central leadership could observe and learn from. The zones worked. The observable success of the zones produced the political support for expanding market mechanisms to broader areas of the economy, which eventually produced the large-scale transformation of the Chinese economy over the subsequent two decades.

The third feature was the partial opening to foreign investment and foreign trade, which exposed Chinese enterprises to external information about production methods, product standards, and customer preferences. The exposure was a resilience-restoring mechanism in the specific sense that it introduced signals from outside the Chinese planning system that could update internal Chinese models. The isolation of the Maoist system had been a contributor to its inability to recognize its own calibration failures. The opening reversed the isolation, at least partially, and allowed external information to begin shaping internal decisions.

What the Deng reforms did not do was fully transition China to a market system. The reforms produced a hybrid system in which market mechanisms coexist with substantial state planning, state ownership, and political control over key sectors. The hybrid has produced sustained growth for more than four decades, which is a remarkable economic achievement, but it has also preserved some of the structural features that produced the Great Leap Forward's failure, in forms that have been moderated but not eliminated. The persistent political risk of leadership-driven campaigns that override local feedback remains visible in the Chinese system, and it is one of the features that distinguishes the Chinese post-collapse construction from the Western post-2008 construction or the post-Soviet transition.

The long lag between the collapse and the reform is the lesson that matters most here. The institutional response to the Great Leap Forward could not begin while the political leadership that had produced the Great Leap Forward remained in power, because the response would have required explicit acknowledgment of the failure that the leadership had produced. The acknowledgment was politically impossible, so the response was deferred, and during the deferral the costs of the collapse were compounded by the costs of the regime's continued operation under the architecture that had produced the collapse. The Cultural Revolution was, in substantial part, the cost of this deferral. Millions of additional deaths, a lost generation of educational and economic development, the persecution of the specific technical and managerial classes whose capacities would have been required for earlier reform, all of these were produced by the political inability to acknowledge the Great Leap Forward's failure directly.

The lag in the Chinese case is instructive because it shows that post-collapse construction is not automatic. It requires political conditions that permit acknowledgment of the failure. When the conditions are absent, the construction is delayed, and the delay extracts substantial additional costs. The costs are not always visible as a second collapse. Sometimes they are visible as a prolonged period of compounded suffering that extends the original collapse rather than producing a new one.

The Shared Shape of Post-Collapse Construction

The four cases show significant variation in their specifics. They also share several features worth making explicit.

The first shared feature is that the construction was reactive. In each case, the institutions that eventually stabilized the post-collapse environment were built in response to the collapse, not in anticipation of it. The Peace of Augsburg was built after thirty-eight years of religious war. Dodd-Frank was passed two years after the 2008 collapse. The post-Soviet market infrastructure was built in the 1990s after the Soviet system had ended. The Deng reforms began more than a decade after the famine. In each case, the preconditions for construction included a visible collapse that had produced consensus, at least within the relevant political elite, that the pre-collapse architecture could not be restored.

The second shared feature is that the construction was partial. None of the post-collapse institutions eliminated the failure mode that the collapse had revealed. They contained it, channeled it, or transformed it into a different kind of failure mode that was more tractable. The modern state contained religious conflict through territorial partition. Dodd-Frank channeled financial risk into regulated structures where it could be monitored. Market infrastructure transformed the planning problem into a distributed computation that many agents could carry out. Deng's reforms transformed the communist planning problem into a hybrid that preserved some of the original failure mode's features while moderating others. The partiality is not a criticism. It is a feature. The post-collapse institutions could not eliminate the underlying cognitive failures that had produced the amplifier in the first place. They could only build mechanisms that prevented the failures from propagating at civilizational scale.

The third shared feature is that the construction was expensive. The direct costs of the collapses themselves, in human lives and in economic destruction, were enormous. The costs of the constructions, in political capital and in institutional experimentation, were also substantial. The post-collapse world was not cheaper than the pre-collapse world. It was a world in which very expensive institutions had been built because the alternative had proven to be more expensive still. The construction costs of each of these cases ran into multiple percentage points of GDP sustained over years, and the human costs of the pre-construction periods ran into millions of deaths in three of the four cases.

The fourth shared feature is that the construction incorporated the amplifier rather than attempting to eliminate it. The modern state operated with the printing press rather than without it. Dodd-Frank operated with securitization rather than without it. The post-Soviet infrastructure operated with complex modern markets rather than with the simpler markets that had preceded the Soviet period. The Deng reforms operated with industrial organization rather than reverting to pre-industrial agrarian systems. In each case, the post-collapse institutions accepted that the amplifier was now a permanent feature of the informational or economic environment and built correction infrastructure that could operate alongside it. The correction infrastructure was what allowed the amplifier's benefits to be retained while its most destructive effects were contained.

The fifth shared feature is that the correction infrastructure operated at throughputs comparable to the throughput of the amplifier it was built to correct. Stress testing operates at the throughput of the modern financial system. The market infrastructure operates at the throughput of the modern economy. The educational system operates at the throughput of modern information production. In each case, the correction was not a lower-throughput brake on the amplifier. It was a higher-throughput feedback mechanism that could catch specific failures as they occurred rather than waiting for them to propagate.

These five features constitute the shape of post-collapse construction. The shape is visible in the historical record, and the architectural pressures that produced it in the previous cases are present in the agent economy case. The remaining question is whether the shape can be produced preemptively, at substantially lower cost than the reactive construction, and only if the construction begins in the window before the first visible failures force a reactive response. The window is still open. It is also closing.

Chapter 11: Building Before the Collapse

Every previous chapter of this book has been diagnostic. This chapter is constructive. The shift is not stylistic. The previous chapters could be written without committing to any specific position on what should be built, because the question of what has happened and what is happening does not depend on the answer to the question of what should be done. This chapter cannot be written the same way. It must propose an architecture, defend the proposal against the most obvious objections, and accept the risk that the proposal is wrong in ways that the next five years will make visible. Proposals in books age. Diagnoses sometimes age better. The choice to close the book with a proposal rather than a diagnosis reflects the judgment that the window for preemptive construction is narrow enough that a proposal, even an imperfect one, is more valuable than a further diagnosis.

The argument of the chapter is specific. A calibration layer for the agent economy is constructible with current technology, at a cost that is a small fraction of the cost of the reactive construction the previous chapter examined, and it can be built in the window between now and the first major visible failure. The construction is not easy, and it is not guaranteed to succeed. It requires choices about architecture, about governance, about commercial model, and about standards, and the choices have to be made in conditions of incomplete information.

But the construction is possible, and the historical analysis of the previous chapters provides a basis for describing what its architecture should look like.

The chapter proceeds in four sections. The first names the specific failure mode the calibration layer would correct and distinguishes it from adjacent problems that the layer would not address. The second describes the architecture of the layer in functional terms, without committing to specific technical implementations, because the technical implementations should follow from the functional specification rather than driving it. The third addresses the governance and commercial questions that any real-world construction would have to answer. The fourth closes the book with the strategic argument for preemptive construction.

The Failure Mode to Correct

The failure mode this book has traced is the one the Substack essay named first: an agent economy that operates without calibration of the premises underneath the queries it processes. The failure is not about whether the agents can execute assigned tasks competently. They frequently can. The failure is not about whether the agents can synthesize information, generate content, or automate workflows at scale. They demonstrably can. The failure is about what happens at the point where the agent receives a query and begins to process it without having any mechanism to assess whether the premise of the query corresponds to the state of the world the query is meant to operate on.

A calibration layer would address this specific failure and not the adjacent failures that other interventions are better suited to correct. It would not address model safety in the sense that alignment research addresses. That work is critical, and it operates on the agent's values and objectives, which is upstream of the calibration question this book has focused on. It would not address model capability limitations in the sense that prompt engineering and fine-tuning address. Those limitations are real, and they matter for specific applications, but they are distinct from the premise-calibration problem. It would not address the regulatory questions that privacy law, copyright law, and sectoral AI regulation address. Those questions are being worked on by specific policy communities, and this book's framework does not replace their work. What the calibration layer would address is the specific failure mode of uncalibrated premises entering the agent stack and propagating through it without any mechanism registering the miscalibration.

The specificity matters for two reasons. It matters because a layer that tries to do everything will do nothing well, and the calibration problem is sharp enough that a dedicated layer has a chance of solving it. And it matters because the commercial model for the layer has to be specifiable. A layer that addresses a specific failure mode can be sold to specific buyers who have specific exposure to the failure, which is what makes its construction economically viable in the window before the first major failures force a regulatory response. A diffuse layer that addresses general concerns about AI is much harder to fund, much harder to position, and much harder to measure.

The specific failure, stated compactly, is the following. Agent queries arrive into the agent stack carrying implicit assumptions about the state of the world that the query addresses. The assumptions include: that the relevant facts are as the query implies, that the relevant actors are in the positions the query assumes, that the relevant environment responds in the ways the query presumes. These assumptions can be more or less correct. When they are substantially incorrect, the agent's output, however well-executed, will be of little value, and may be positively harmful if it is acted on. The calibration layer's function is to evaluate the assumptions of the query against the state of the world the query addresses, to flag the discrepancies, and to route the query through a correction process before the agent proceeds to execute.

The Architecture

The layer has four functional components, each corresponding to one of the four dimensions the book has been using as diagnostic vocabulary. The four components operate on every query that passes through the layer, producing assessments that can be combined to produce an overall calibration judgment or that can be used independently depending on the specific deployment context.

The first component is the calibration check. Its function is to evaluate whether the factual premises of the query correspond to the state of the world the query is operating on. For a query that asks the agent to do something based on data about a customer, a product, a market, or a process, the calibration check would evaluate whether the data referenced in the query matches the current state of the organization's authoritative data sources on the same subjects. The check would not require complete certainty, which is impossible. It would produce a confidence assessment, a flag when the confidence falls below a specified threshold, and a record of the specific discrepancies that produced the flag. The flag would trigger human review or automated re-routing depending on the deployment configuration.

The calibration check requires access to the organization's authoritative data sources, which is a non-trivial engineering problem but is not a conceptually novel one. Enterprise data integration has been a mature engineering discipline for decades. The calibration check does not require new data infrastructure. It requires the specific integration that allows the layer to query the organization's data at the point of agent query processing, and to do so at latencies that the agent query workflow can tolerate.

The second component is the navigation check. Its function is to evaluate whether the workflow the query is about to execute is the appropriate workflow for the query's actual purpose, given the calibrated premises. An agent query that arrives with one stated purpose may, after the premises are recalibrated, turn out to have a different actual purpose that should be pursued through a different workflow. The navigation check is the function that catches this kind of misrouting. It evaluates the congruence between the query's stated intent, the calibrated factual state, and the workflow that would execute on the query. When the three are not congruent, the navigation check flags the misrouting and routes the query back to a higher-level orchestration function that can reassess.

The navigation check is less developed in current enterprise AI infrastructure than the calibration check, because the concept of orchestration is still being defined as the industry moves from single-agent deployments to multi-agent workflows. The navigation check's specification depends on the orchestration architecture being built above it, which means that the check has to be designed to be adaptable to different orchestration approaches rather than tightly coupled to a specific one. The adaptability is achievable through a clean functional interface: the navigation check receives the query, the calibrated premises, and the proposed workflow, and returns a congruence assessment.

The third component is the resilience check. Its function is to evaluate whether the query's premises and workflow remain valid in conditions that the training and testing environment may not have anticipated. The resilience check is the layer's defense against the failure mode of environments that shift faster than the agent's internal model can update. For each query, the check evaluates whether the current environment matches the environment in which the agent was trained and validated, and flags discrepancies that suggest the agent's behavior may not generalize to the current context. The check does not require the agent to adapt in real time. It requires the layer to surface the generalization risk so that the deploying organization can decide how to handle it.

The resilience check is the most technically ambitious of the four components, because it requires the layer to maintain its own representation of the environment the agent is operating in, and to track drift in that environment over time. The technical approaches to this are active areas of research: distribution shift detection, concept drift monitoring, and out-of-distribution input detection are all mature fields whose techniques can be adapted for the resilience check. The adaptation is non-trivial, but it is not blocked by fundamental research questions. It is an engineering problem with a specification that research can support.

The fourth component is the outcome achievement check. Its function is to evaluate whether the outputs the agent is about to produce can actually be converted into the outcomes the organization deployed the agent to achieve. The check is about the bridge from output to outcome, which is the bridge that Chapter 2 identified as the most frequently broken in institutional deployments. The check evaluates whether the output format, the output delivery mechanism, the downstream processes that will consume the output, and the measurement systems that will verify the outcome are all in place and functional. When they are not, the check flags the gap and routes the query to a completion process that addresses the gap before the agent proceeds.

The outcome achievement check is the component most likely to produce immediate value for deploying organizations, because it addresses a frequent source of agent deployment disappointment: workflows that produce outputs that cannot be used. The check's value is demonstrable in short time horizons, which makes it a good entry point for organizations that are evaluating whether a calibration layer is worth adopting. The entry point consideration matters for the commercial viability of the layer, because the layer's adoption will be gradual, and early adopters will choose based on the component that produces the most immediate measurable value.

The four components combine into a layer that sits between the query generation function and the agent stack. Every query passes through the layer. Some queries pass through without any flags, because their premises are well-calibrated, their workflows are appropriate, their environments are consistent with training, and their outcome conversion paths are functional. Those queries proceed to the agent stack and are executed normally. Some queries produce flags in one or more components, which triggers the routing to human review or automated correction. The throughput of the layer is designed to match the throughput of the agent stack it is protecting, which is high, because the agent stack operates at machine speed and the layer must not become a throughput bottleneck.

The architecture described is deliberately abstract. It does not specify the specific technologies that would implement each component, because the specific technologies will depend on the deploying organization's existing infrastructure, on the capabilities of the agent stack being protected, and on the specific industry context in which the layer is operating. What the architecture does specify is the functional requirements: every agent query passes through four checks, each check produces a specific kind of assessment, the assessments combine to produce routing decisions, and the routing decisions operate at throughput that matches the agent stack. Different implementations of this architecture are possible. The argument is that the architecture itself is the construction that the agent economy needs.

Governance and Commercial Viability

The architecture alone is insufficient. A calibration layer that is technically specified but has no governance framework and no commercial model will not be built at scale. The three questions that any real-world construction would have to answer are: who operates the layer, how is it funded, and what standards does it operate under.

On the operation question, the layer could be operated by each deploying organization independently, by specialized third-party providers, by agent vendors as part of their stack, or by some combination. Each option has trade-offs. Internal operation gives the deploying organization full control over the layer but requires the organization to develop substantial capability in calibration engineering, which is a specialized discipline that few organizations have cultivated. Third-party operation concentrates the capability in specialized providers who can amortize the development costs across many clients, but raises questions about the providers' independence from the agent vendors whose output the providers are evaluating.

Vendor operation creates the cleanest integration between the layer and the stack, but puts the vendor in the position of evaluating its own output, which is a structural conflict of interest that the financial audit industry has spent decades learning to manage through specific governance architectures.

The most likely stable equilibrium, based on the history of adjacent industries, is a hybrid in which deploying organizations operate portions of the layer internally, third-party providers operate other portions, and industry standards govern the interfaces between the components. The hybrid is similar to how cybersecurity is currently organized in large enterprises: some functions are operated internally by the enterprise's security team, some functions are outsourced to specialized managed security services providers, and industry standards like ISO 27001 and SOC 2 govern the interfaces. The cybersecurity parallel is instructive because the structural similarities are deep. Both are specialized protective functions that are adjacent to but distinct from the primary business operations, both require substantial specialized expertise, and both produce measurable risk reduction that can be quantified for commercial purposes.

On the funding question, the layer has to be funded by someone who benefits from its existence, and the question is who that is. The deploying organizations benefit, because the layer reduces the risk of agent deployments producing material losses. The insurance carriers benefit, because the layer reduces the expected claims against policies that cover AI-related losses. The agent vendors benefit indirectly, because a calibration layer that prevents the most severe failures preserves the commercial viability of agent deployment in the long run, even if it reduces deployment throughput in the short run. The regulators benefit, because a functioning calibration layer reduces the frequency of incidents that force regulatory response. The public benefits, because the most severe failure modes are the ones that produce the broadest external costs.

The most viable short-term funding sources are the deploying organizations and the insurance carriers. The deploying organizations can be sold the layer as a risk management investment, with the commercial case supported by the legal exposure that agent deployments are already beginning to produce. The insurance carriers can support the layer through premium discounts for deployments that include calibration infrastructure, which creates an economic incentive for deploying organizations to adopt the layer even when the organizations have not yet made an internal judgment about the necessity. Both funding mechanisms are already being explored at the edges of the market, and the scaling question is whether the adoption curve accelerates fast enough to produce a functioning layer before the first major failures force reactive construction.

On the standards question, the layer needs an interoperable specification that allows different implementations to be compared, evaluated, and integrated. The specification work is the role of standards bodies, whose slow pace is a well-known frustration but whose outputs are what makes cross-organization interoperability possible. For the calibration layer, the specification work would need to address: what constitutes an adequate calibration check, what constitutes an adequate navigation check, what constitutes an adequate resilience check, what constitutes an adequate outcome achievement check, and how the four combine to produce a defensible overall calibration judgment. The specification work is substantial but not unprecedented. Financial audit standards, medical device standards, and safety-critical software standards all provide templates that could be adapted for the calibration layer.

The standards work should begin now, even though its outputs will be years in arriving. Standards that arrive after the failure mode has been widely recognized are typically rushed and shaped by the specific failures that produced the recognition, which means they are optimized for the past rather than for the present. Standards that are developed in the window before the major failures have the opportunity to be shaped by the architectural analysis rather than by incident response. The standards developed in this window will be imperfect, but they will be more thoughtful than the ones that would be developed under the pressure of a visible collapse.

Why the Preemptive Construction Matters

The argument for preemptive rather than reactive construction is strategic, and its components should be named explicitly before the book closes.

The first component is the cost differential between preemptive and reactive construction. The historical record of the previous chapter shows that reactive construction has consistently run into costs that are multiples of what preemptive construction would have cost, had preemptive construction been feasible. The multiples are not small. They are often the difference between an expense a society can absorb and an expense that consumes a generation's capacity for other development. The agent economy case is not exactly parallel, because the agent economy is not yet at the scale where its failure would consume a generation. But the cost differential is likely to hold, because the structural pressures that produced the differential in the previous cases are present in the current case as well.

The second component is the optionality that preemptive construction preserves. A calibration layer built before the first major failure preserves the option to continue deploying agents at throughput and to continue realizing the productivity benefits that agent deployment is producing. A calibration layer built after the first major failures will likely be built under regulatory constraint, and the constraints will reduce the throughput and limit the benefits. The optionality argument is about the state of the commercial and regulatory environment in which the calibration infrastructure will be built, not about the technical features of the calibration infrastructure itself. The same technical features, built in the pre-failure environment, are a competitive advantage; built in the post-failure environment, they are a regulatory requirement. The economic value of the same capability differs significantly between the two environments, and the difference is captured by whoever is holding the capability at the right moment.

The third component is the informational advantage of preemptive construction. A calibration layer built now can be built with attention to the specific structural features of the agent stack as it currently exists, and can be adapted as the stack evolves. A calibration layer built after the failure will be built with attention to the specific failures that produced the reactive pressure, which will optimize the layer for preventing the last failure rather than for catching the next one. The pattern is visible in the post-2008 construction, which is exceptionally well-optimized for preventing another 2008-style collapse and which is less well-optimized for preventing the failures that financial regulators are currently concerned about. The informational advantage of preemptive construction is that the construction can be shaped by the architectural analysis, which generalizes, rather than by the incident analysis, which does not.

The fourth component is the civilizational stakes. The agent economy is currently being deployed into the specific decisions that shape how a society allocates its resources, selects its talent, treats its sick, adjudicates its disputes, and runs its operations. The failures, when they arrive, will arrive in these domains. The scale at which they will arrive is not yet fully visible, because the deployments are still expanding, and the failures that will occur at full deployment have not yet occurred at partial deployment. The ratio of decisions-made-through-agents to decisions-made-through-human-review is currently transitioning from low to high, and the failures' visibility will transition with it. The window to build correction infrastructure before the failures force the issue is a window in which the correction can be designed thoughtfully rather than imposed reactively. Thoughtful correction is better correction. Better correction means that the agent economy produces its productivity benefits with a smaller tax in the form of avoidable failures, which is to say that the society captures more of the benefit and pays less of the cost.

Stated compactly: civilizations that build their correction systems before the collapse absorb the failure mode at a cost they can afford. Civilizations that build after, pay the cost of the collapse and then pay the cost of the correction. The question the agent economy poses is which kind of civilization we become. The answer is not predetermined. It depends on whether the window for preemptive construction is used.

The window will remain open for eighteen to thirty-six months. During that time the flags are accumulating in internal systems and the first visible failures are approaching. At some point within the window, the first major failure will arrive and be read as a one-off, a second will arrive and begin to suggest a pattern, and a third will force the pattern into visibility. After the third, the construction will be reactive. Before the first, the construction can still be preemptive.

The failure mode is not new. It is the oldest failure mode humans produce at scale. The architecture for containing it is not new either. It is the architecture previous collapses have eventually produced, at great cost, after the fact. What this moment offers is the opportunity to build the architecture ahead of the cost. That opportunity was never taken in the previous cases, because the pattern was not yet legible. It is legible now.

The nervous system without a mind is not permanent. It becomes permanent only if the mind is not built. The mind is buildable. The window is open. The work begins when the book closes.

Appendices

Appendix A: The Amplifier Audit

The twelve questions below are designed to allow any organization to make a rapid self-assessment of whether it is operating inside an amplifier system. The questions are not exhaustive. They are the set that experience suggests produce the sharpest signal in the shortest time. An organization that answers no to most of them is probably not operating inside an amplifier in a material way. An organization that answers yes to most of them is almost certainly operating inside one, and should begin the more detailed assessment that Chapter 11's architecture supports.

The questions are grouped by the four dimensions of the Risk Efficacy framework, with three questions per dimension. Each question can be answered yes, no, or uncertain, with uncertain treated as yes for auditing purposes. The bias toward conservatism reflects the practical reality that uncertainty about whether a correction system is in place typically means the correction system is not in place.

Calibration

1. Does the organization have a documented mechanism by which the premises underneath automated decisions are validated against

current authoritative data, and does the mechanism operate at the throughput of the decisions it validates?

2. When an automated system produces an output that downstream review identifies as miscalibrated, does the organization have a process that traces the miscalibration back to the specific premise that was wrong, records the correction, and incorporates the correction into subsequent decisions?

3. If the organization were asked to demonstrate, for a specific automated decision made in the last thirty days, that the premises on which the decision rested corresponded to the actual state of the world at the time of the decision, could it produce the demonstration with evidence?

Navigation

4. For workflows that include automated steps, does the organization have explicit checkpoints at which a human reviews whether the workflow the automation is executing is the appropriate workflow for the actual situation, rather than the workflow the automation was configured to execute?

5. When an automated workflow produces outputs that get combined with outputs from other workflows to produce downstream actions, is there a function that evaluates whether the combined outputs are coherent and whether the action they support is the action the organization intended?

6. Can the organization identify, for its major automated workflows, the specific senior individuals who are accountable for the workflows' outcomes, and can those individuals describe what they would do if the workflows began producing results that diverged from expectations?

Resilience

7. Does the organization monitor the environment in which its automated systems operate for drift from the conditions under which the systems were designed and validated, and does the monitoring produce alerts that trigger specific human review?

8. When external conditions change in ways that could affect the validity of automated decisions, does the organization have a protocol for assessing the impact of the change on the automated systems and for pausing or adjusting the systems if the impact warrants?

9. If the assumptions underlying the organization's major automated systems were invalidated by a significant external event, does the organization have an inventory of which systems would be affected and which decisions would need to be reevaluated?

Outcome Achievement

10. For major automated workflows, does the organization have a documented specification of what outcomes the workflow is meant to produce, separate from the outputs the workflow directly generates?

11. Is there a measurement system that tracks whether the workflow's outputs actually convert into the specified outcomes, and does the measurement system produce regular reports to the responsible parties?

12. When the measurement system shows that outputs are not converting into outcomes, is there a process that investigates the divergence, identifies its cause, and triggers adjustments to the workflow?

Scoring

The scoring is deliberately crude. Count the number of questions answered yes. Zero to three indicates an organization that has not yet developed meaningful correction infrastructure and is exposed to the failure mode at close to the theoretical maximum for its current deployment intensity. Four to six indicates an organization with partial correction infrastructure, typically concentrated in specific functions or domains, with substantial gaps elsewhere. Seven to nine indicates an organization with developed correction infrastructure across most dimensions, with specific gaps that can be identified and addressed. Ten to twelve indicates an organization that has built the architecture Chapter 11 describes, or one substantially similar to it, and that is operating at the current frontier of calibration practice.

The scoring is not a grade. Organizations in the zero-to-three range are not necessarily failing. They may be at earlier stages of automation deployment, or may be operating in domains where the amplifier architecture has not yet reached the intensity that would produce the failure mode. The scoring is a rough indicator of where the organization currently sits on the trajectory this book has described, and of how much construction work is required to move toward the architecture that will be necessary when the agent economy reaches its full deployment intensity.

The audit is intended to be done in a single sitting, by a small group of senior leaders who have cross-functional visibility into the organization's operations. It should not be delegated to automation specialists, because the audit is about whether the specialists' work is integrated with the rest of the organization's decision infrastructure, which is a question the specialists cannot answer for themselves. A single audit produces a snapshot. Repeated audits, done quarterly or semi-annually, produce a trajectory, which is the information that is actually useful for planning construction.

Appendix B: Historical Sources

The historical chapters of this book draw on substantial secondary literature in several fields. This appendix names the principal works that shaped the arguments. It is not a full bibliography. It is the short list of texts that a reader who wants to pursue the historical arguments in their own right should consider first. Each chapter's note indicates the specific works that were most directly relevant to the chapter's framing.

Introduction and Chapter 1

The framing of the agent economy draws on the contemporary technical literature on large language models, retrieval-augmented generation, and multi-agent systems. The underlying measurement framework is developed in Wu, A. (2026), *Risk Efficacy: How to Measure Decision Making in an Age of Uncertainty.* Readers interested in the distinction between the expert marketplace thesis and the calibration thesis may also consult Wu, A. (2026), *The Neurodivergent Advantage,* which applies the framework to a specific cognitive population.

Chapter 2

The four-dimensional structure of the Risk Efficacy framework is specified in Wu, A. (2026), *Risk Efficacy: How to Measure Decision Mak-*

ing in an Age of Uncertainty. The extension of the framework to institutional and infrastructural scales, which this book undertakes, draws on work in organizational theory, decision science, and cognitive psychology. Relevant background reading includes Daniel Kahneman's *Thinking, Fast and Slow* (2011) on cognitive biases at the individual scale, and Charles Perrow's *Normal Accidents* (1984) on failure modes at the institutional scale.

Chapter 3

The historical literature on the Reformation is vast. The most directly relevant texts for this book's argument are Elizabeth Eisenstein's *The Printing Press as an Agent of Change* (1979), which develops the thesis that the printing press was not merely the medium through which the Reformation spread but was itself a structural condition for the Reformation's occurrence. The economic history of indulgences is developed in detail in Berndt Hamm's work on late medieval piety. For the political and ecclesiastical response, Diarmaid MacCulloch's *The Reformation* (2003) provides a comprehensive account, and for the long-term institutional consequences, the scholarship on early modern state formation, including works by Charles Tilly and James Scott, establishes the framework this chapter draws on for the "modern state" construction.

Chapter 4

The historical literature on the Great Leap Forward has expanded significantly since the opening of Chinese archives in the 1980s. The most important works include Frank Dikötter's *Mao's Great Famine* (2010), Yang Jisheng's *Tombstone* (2012), and Judith Banister's demographic work on the mortality figures. The information-flow analysis of the reporting chain draws on Carl Riskin's work on the Chinese economy and on the detailed case studies collected in the Chinese-language scholarship on provincial and county records from the period.

The subsequent reform period is treated in Ezra Vogel's *Deng Xiaoping and the Transformation of China* (2011) and in the extensive literature on Chinese economic reform.

Chapter 5

The literature on the 2008 financial crisis is enormous and varies significantly in analytical quality. The most rigorous accounts for the purposes of this book's argument are Gary Gorton's *Misunderstanding Financial Crises* (2012) and *Slapped by the Invisible Hand* (2010), which develop the structural analysis of how securitization transformed the information architecture of credit markets. Anat Admati and Martin Hellwig's *The Bankers' New Clothes* (2013) develops the policy implications. The Financial Crisis Inquiry Commission's final report (2011) remains the most comprehensive official documentary source. For the regulatory response, the detailed literature on Dodd-Frank implementation, including scholarship by Daniel Tarullo from his tenure on the Federal Reserve Board, provides the authoritative account.

Chapter 6

The literature on Soviet central planning and its collapse includes both contemporaneous and retrospective scholarship of varying quality. The most analytically rigorous works are Alec Nove's *An Economic History of the USSR* (1969, with later editions), Janos Kornai's *The Socialist System* (1992), which develops the analytical framework this chapter draws on, and Yegor Gaidar's *Collapse of an Empire* (2007), which develops the specific account of the terminal phase. The post-collapse transitions are treated in the detailed comparative literature, including Anders Åslund's work on the Russian transition and Leszek Balcerowicz's own account of the Polish reform program.

Chapters 7 through 9

These chapters draw primarily on contemporary sources in AI research, enterprise technology deployment, and the early literature on AI governance. Relevant technical references include the ongoing work published through the major AI conferences (NeurIPS, ICML, ICLR) and through the Alignment Research Center's output. Policy-oriented references include the NIST AI Risk Management Framework (2023) and the work of the Center for Security and Emerging Technology at Georgetown. The specific projections in Chapter 9 are the author's, developed against the pattern the earlier historical chapters establish.

Chapters 10 and 11

Chapter 10 synthesizes across the historical sources named above. Chapter 11 proposes constructions that do not yet exist in the literature in the form the chapter specifies, which means the chapter's arguments should be evaluated on their internal coherence and their fit with the diagnostic analysis developed in the earlier chapters rather than on their correspondence to published research.

A Note on Method

This book's historical arguments are the author's synthesis of secondary literature rather than original primary research. The book is not a history of the cases it discusses. It is a diagnostic argument that uses the historical cases as evidence for a structural claim about the agent economy. The sources named above are the ones a reader should consult to verify the accuracy of the historical characterizations and to pursue the cases in greater depth than this book's argument permits. Where the cases are treated in ways that existing scholarship would complicate, the complication typically does not affect the diagnostic argument, because the diagnostic argument operates at a structural level that is robust to most of the interpretive debates within the specialized scholarship. Readers who want to engage those debates should begin with the texts named here and follow the footnote chains into the more specialized literature.

Reference List

The references below are organized alphabetically by author. They include every source cited in the body of the book, together with the principal works on which the historical chapters draw. Appendix B provides chapter-by-chapter notes on how these sources informed the arguments.

Admati, Anat, and Martin Hellwig. 2013. The Bankers' New Clothes: What's Wrong with Banking and What to Do about It. Princeton: Princeton University Press.

Åslund, Anders. 2007. *How Capitalism Was Built: The Transformation of Central and Eastern Europe, Russia, the Caucasus, and Central Asia.* Cambridge: Cambridge University Press.

Balcerowicz, Leszek. 1995. *Socialism, Capitalism, Transformation.* Budapest: Central European University Press.

Banister, Judith. 1987. *China's Changing Population.* Stanford: Stanford University Press.

Dikötter, Frank. 2010. *Mao's Great Famine: The History of China's Most Devastating Catastrophe, 1958–1962.* New York: Walker & Company.

Eisenstein, Elizabeth L. 1979. *The Printing Press as an Agent of Change: Communications and Cultural Transformations in Early Modern Europe.* Cambridge: Cambridge University Press.

Financial Crisis Inquiry Commission. 2011. *The Financial Crisis Inquiry Report: Final Report of the National Commission on the Causes of the Financial and Economic Crisis in the United States.* Washington, D.C.: U.S. Government Printing Office.

Gaidar, Yegor. 2007. *Collapse of an Empire: Lessons for Modern Russia.* Washington, D.C.: Brookings Institution Press.

Gorton, Gary B. 2010. *Slapped by the Invisible Hand: The Panic of 2007.* New York: Oxford University Press.

Gorton, Gary B. 2012. Misunderstanding Financial Crises: Why We Don't See Them Coming. New York: Oxford University Press.

Hamm, Berndt. 2004. *The Reformation of Faith in the Context of Late Medieval Theology and Piety.* Edited by Robert J. Bast. Leiden: Brill.

Kahneman, Daniel. 2011. Thinking, Fast and Slow. New York: Farrar, Straus and Giroux.

Kornai, János. 1992. The Socialist System: The Political Economy of Communism. Princeton: Princeton University Press.

MacCulloch, Diarmaid. 2003. *The Reformation: A History.* London: Allen Lane.

National Institute of Standards and Technology. 2023. *Artificial Intelligence Risk Management Framework (AI RMF 1.0).* NIST AI 100-1. Gaithersburg, MD: U.S. Department of Commerce.

Nove, Alec. 1992. *An Economic History of the USSR, 1917–1991.* 3rd ed. London: Penguin Books. [First edition 1969.]

Perrow, Charles. 1984. *Normal Accidents: Living with High-Risk Technologies.* New York: Basic Books.

Riskin, Carl. 1987. China's Political Economy: The Quest for Development since 1949. Oxford: Oxford University Press.

Scott, James C. 1998. *Seeing Like a State: How Certain Schemes to Improve the Human Condition Have Failed.* New Haven: Yale University Press.

Tarullo, Daniel K. 2019. "Financial Regulation: Still Unsettled a Decade after the Crisis." *Journal of Economic Perspectives* 33 (1): 61–80.

Tilly, Charles. 1990. *Coercion, Capital, and European States, AD 990–1992.* Cambridge, MA: Blackwell.

Vogel, Ezra F. 2011. *Deng Xiaoping and the Transformation of China.* Cambridge, MA: Belknap Press of Harvard University Press.

Wu, Alia. 2026a. *Risk Efficacy: How to Measure Decision Making in an Age of Uncertainty*. New York: Redline Rising.

Wu, Alia. 2026b. *The Neurodivergent Advantage: Risk Efficacy and the Measurement of Different Minds*. New York: Redline Rising.

Yang, Jisheng. 2012. Tombstone: The Great Chinese Famine, 1958–1962. Translated by Stacy Mosher and Guo Jian. New York: Farrar, Straus and Giroux.

ABOUT THE AUTHOR

Alia Wu is an applied cognitive scientist and founder of Risk Efficacy. Her work spans clinical neuroscience, quantitative finance, and professional motorsport. She builds measurement systems for decision quality in high-stakes environments. Wu is also President of Redline Rising, a nonprofit research and training platform focused on agency training and measurement. *The Amplifier* is a companion volume to her Meritocracy Society Infrastructure Series.

www.ingramcontent.com/pod-product-compliance
Lightning Source LLC
LaVergne TN
LVHW022218200926
848481LV00006B/2015
* 9 7 9 8 9 9 4 5 2 4 2 5 1 *